HARNESSING WATER POWER FOR HOME ENERGY

by Dermot McGuigan

Garden Way Publishing Co.

Charlotte, Vermont   05445

Printed in the United States by Capital City Press
Designed by Bruce Williamson
Illustrations by Cathy Baker
Cover design by Trezzo/Braren Studio

McGuigan, Dermot, 1949–
    Harnessing water power for home energy

    Bibliography: p.
    1.   Water-power electric plants. I. Title.
TK1081.M26   621.312134   77-27404
ISBN 0-88266-115-9 pbk.

Contents

INTRODUCTION

Over the last few years there has been a resurgence of interest in small-scale water power, and there are available, right now, whole new ranges of water turbines and power systems for home use. In fact there are systems to suit just about every pocket, from basic low-energy DC systems to fully governed automatic AC systems. Moreover, there are turbines to fit every set of conditions, from small mountain streams to wide valley rivers.

It is true to say that most people with flowing water on their land can now begin to benefit appreciably from that source of power, for there is a great deal of exciting activity going on in the water power field today. This book takes a close look at what is happening.

In the past, the miller would build a mill-pond to store water during the night for use by the watermill during the day. This involved flooding good land, building a dam or bank, providing an overflow, and a basic flow regulator. In other words it involved a lot of money and hard work. With the arrival of modern electricity-producing turbines, in the late 19th Century, a new problem arose. Power generated by the turbine had to be exactly the same as power required. This involved building turbines with regulators which could control, to an exact degree, the flow of water into the turbine, thus closely regulating the output to match the need. In many cases the regulating device or governor cost more than the turbine itself!

Modern water power developers have now overcome these problems. On small domestic turbine plants, batteries can now be used to store electricity on a daily basis and thus replace the old mill-pond. This is suitable on turbines with outputs of .5 to 3 kW (500 to 3000 watts or more). Inverters may be used to give utility-type alternating current (AC) electricity, or better still it is possible to use a direct current (DC) system, thus avoiding inverter costs. Turbines on a battery system do not require complicated governors.

A recently developed electronic device called the Gemini Synchronous inverter makes it possible to use the utility lines as a mill-pond! The Gemini inverter is interfaced between the utility line and the turbine generator.

There are many advantages to this system: The turbine does not need a governor for close speed regulation. The ungoverned output from the generator is instantaneously converted to utility-type electricity, which can then be used in the home. If the power company is willing, their line can be used as a battery or energy reservoir. The turbine feeds power into the line when it is not required and draws it out when it is.

This system can only be used next to a utility line, and its single disadvantage is its dependence upon the power company. But if it goes out on strike or runs out of fossil fuel to burn, then a battery system or some other regulating device can be used.

There are now available two new types of governors which regulate the electrical output rather than govern the flow of water. The first of these is the electronic load diversion governor. The diversion governor is a fairly simple device which diverts the alternator output to and from the primary load to a secondary load as and when needed.

The second type of electronic alternator governor regulates the alternator output to suit varying electric loads by changing the charge to the alternator field. In other words, even at full turbine output, the current from the alternator will be exactly matched to the load. Got it? If not, don't worry, as we will be looking at all these methods later on.

Finally, there is another relatively uncomplicated system suitable for use with many turbines. This involves a mechanical water flow diverter governor. The diverter governor is a simple way of regulating the amount of water striking the turbine, thus matching output with demand.

Small-scale water power is not going to solve all our energy problems. But there are a great number of people who own sites suitable for the development of water power, and it is only sensible that this power should be used. In many cases the available water power is sufficient to meet all the energy needs of the home or farm, but if not, solar energy, wood burning or wind power can be used as back-ups. Those connected to the public utility can use it as back up.

Throughout this book, I have concentrated on the generation of electricity from water power. This is because electricity can be made to do a great variety of work, but it is worth recalling the numerous uses to which water power was put in the past. The monastic orders of Europe made great use of it. The following account of a watermill, which was once at Clairvaux, is translated from Abbe Vacandard's 'Vie de St. Bernard':

"Entering the abbey under the boundary wall, the stream first hurls itself impetuously at the mill where in a welter of movement it strains itself, first to crush the wheat beneath the weight of the millstones, then to shake the fine sieve which separates flour from bran.

Already it has reached the next building: it replenishes the vats and surrenders itself to the flames which heat it up to prepare beer for the monks, their liquor when the vines reward the wine-growers' toil with a barren crop.

The stream does not yet consider itself discharged. The fullers established near the mill beckon to it. In the mill it had been occupied in preparing food for the brethren: it is therefore only right that it should now look to their clothing. It never shrinks back or refuses to do anything that is asked of it. One by one it lifts and drops the heavy pestles, the fullers' great wooden hammers. When it has spun the shaft as fast as any wheel can move, it disappears in a foaming frenzy; one might say it had itself been ground in the mill.

Leaving here it enters the tannery, where in preparing the leather for the shoes of the monks it exercises as much exertion as diligence; then it dissolves in a host of streamlets and proceeds along its appointed course to the duties laid down for it, looking out all the time for affairs requiring its attention, whatever they might be, such as cooking, sieving, turning, grinding, watering or washing, never refusing its assistance in any task."

Sadly that monastery has long gone, but I have visited a Cistercian monastery, Mount Mellary Abbey, situated on the side of a mountain in County Waterford, Ireland. There, since 1912, they have used both a Francis and a Pelton turbine. From 1912 to 1964 the Francis produced all the electricity required throughout the sizeable Abbey. The Francis continues to supply electricity but in

1964 demand exceeded supply, and the mains won. The Pelton turbine is more interesting in that it is used solely for mechanical power. Today it still mixes the dough for bread in the bakery. In the laundry it drives two machines, and it operates one of the saws in the sawmill. Until recently it also operated the butter churn.

The overflow from the one and a half million gallon reservoir at the Abbey is directed through a series of water gardens and finally operates the organ — yes, a water-powered church organ! It is heartening to see water put to such inventive and aesthetic uses, especially in this day and age where water is given no respect and is frequently polluted to an excessive degree. Man has yet to realize that polluting water damages all life, for life depends upon water.

Whilst there is much in the Chinese way of life that I find objectionable, there is one aspect of their society which I feel the West could learn from; and that is their attitude toward small hydropower. The first Chinese campaign to dot the countryside with small water power plants suitable for electricity generation, irrigation and flood control started in 1956. Guidelines governing the development of small hydro stations are based on the use of local labor and materials. Traditional, indigenous methods of dam construction are employed, and, most dams are either earth-filled or rock-filled structures, so consumption of steel and cement is minimal. Much of the hydroelectric equipment is made in local communal workshops. By 1975 a staggering total of 60,000 installations were completed, with an average output of 35 kW per station. That represents a total capacity of 2.1 million kilowatts. Moreover, the Tientsin Electro-Driving Research Institute is trial-producing seven types of miniature water turbine generators with outputs ranging from 0.6 to 12 kW, suitable for isolated mountain regions.

The potential in the United States for the development of hydroelectric power, large and small, is enormous. The Federal Power Commission has calculated the nation's undeveloped water power resources at an annual 470 billion kilowatt hours, the energy equivalent of burning 100 million tons of oil at a yearly cost of $7 to $10 billion.

In fact, I think the figure is a little higher than this, as many small streams and rivers are not included in the estimate. True, the power potential of these sources is not significant on a national scale, but

on a human scale it can be of great importance to those who benefit directly from them.

This potential may have been better understood back in 1880,
when it has been estimated there were 9,700 water-wheels operating
in New York State alone. Today only a fraction of that power continues to be used. The rest runs to waste.

There is much to commend water power. Simple operation,
trouble- and pollution-free running, with minimal maintenance
costs and using a limitless natural energy source, all combine to
make the idea very attractive. It is probable that within 5 or 10 years
the owners of small hydroelectric plants will be regarded with envy
by neighbors whose fuel bills are soaring, or who are being severely
rationed. Perhaps the real beauty of hydropower is that with a little
care and attention it will go on serving, if not indefinitely, then
certainly for a long time.

In writing this book, I was brought into contact with many who
installed their own turbines, and without exception I can say that
they numbered among the finest men I have ever met. They produce
their own power, and should something go wrong they rarely go
off and complain to a third party but rather, having the ability to
fix it themselves, they do so. It is an independence I greatly admire,
an independence born of enthusiasm and interest to make use of an
excellent and renewable energy source. In particular, I would like to
thank the following who have all in their own ways contributed
to this book:

BILL DELPH
RUPERT ARMSTRONG EVANS
WILLIAM KITCHING.
FRANCIS SOLTIS
LORD WILSON, C.ENG.
JOHN WOOD
DR. MORRIS WRIGHT

How To Measure
Flow and Head

To determine feasibility for developing a water power system, this is where you must start.

This is a job which takes time and care. Its importance is based on the fact that a lot of money may be invested on the basis of the figures. In order to gain an understanding of a stream, one should really spend time with it, absorbing its detail, watching it flow, so that the most appropriate site for the installation can be found. Ideally the flow of the potential power source should be measured for a year before any hardware is purchased. If this is done, then two important facts are known:

1. The lowest and highest seasonal flows — the extremes.
2. The average, dependable flow per month.

With a firm knowledge of the flow figures and the available head, then a turbine can be installed with the certainty that it will produce a given amount of power each month.

The purpose for which the turbine is being installed should then be examined. For example, if it is to supply domestic power requirements, then how much power is needed and when? When you know what the river can supply, and the domestic demand, you can then plan accordingly.

The power in any stream is purely a function of the available flow (Q) and the head (H). Therefore power (P) is equal to $Q \times H$. Some people are fortunate in that a local water authority has kept detailed readings of the river flow, but for those who are not so lucky there are three ways to find the value of Q.

Flow Measurement

Container Method

This method is only suitable for small mountain streams. Build a dam, divert the whole stream into a container of known size and time how long it takes to fill. It may not be necessary to build a dam, a groove could be built into the stream bed where the flow gathers at the point of a fall and a pipe inserted into the groove to draw off the flow into the container.

> **Example:**   *A 60 gallon tank is used, and it takes 30 seconds to fill. Therefore Q equals 120 gallons a minute, which is equal to 10 cubic feet per minute (cfm).*

Weir Method

This is the most accurate method of measuring the flow in medium sized streams. The weir is built like a dam across the stream, which causes all the water to flow over a rectangular notch of known dimensions, see figure below. The weir is best constructed with timber and made watertight with sandbags, sods or clay.

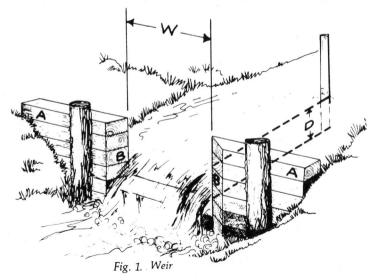

Fig. 1. Weir

Before building the weir, take an approximate measurement of the stream, and see that the overflow notch is sufficient to take maximum flow. The notch should have a width to height ratio of at least 3:1, and be perfectly level and sharp-edged.

To measure the depth of water flowing over the weir, drive a stake in the stream bed three or more feet upstream from the weir, to a depth such that a mark on the stake is exactly level with the bottom of notch "B." Measure the depth "D" in inches of water over the mark, and read the volume of flow in cubic feet per minute per inch of notch width from the table below. Multiply this volume by the notch width in inches, to obtain the total stream flow in cubic feet per minute.

WEIR TABLE

Depth on stake in inches.	Cubic ft. per min. per inch length.	Depth on stake in inches.	Cubic ft. per min. per inch length.
1	0.4	10	12.7
1.5	0.7	10.5	13.7
2	1.1	11	14.6
2.5	1.6	11.5	15.6
3	2.1	12.5	16.7
3.5	2.6	12.5	17.7
4	3.2	13	18.8
4.5	3.8	13.5	19.9
5	4.5	14	21.1
5.5	5.2	14.5	22.1
6	5.9	15	23.3
6.5	6.6	15.5	24.5
7	7.4	16	25.7
7.5	8.2	16.5	26.9
8	9.1	17	28.1
8.5	10.0	17.5	29.4
9	10.8	18	30.6
9.5	11.7	18.5	31.9

Example: *A weir is 3 ft. 6 in. wide and the depth of water at the stake is 10 inches. The flow in cubic feet per minute is therefore 42 × 12.7 = 533 cfm. Once the weir is constructed (easier said than done) it is a simple matter to take frequent readings.*

Float Method

This method is not as accurate as the two above. The flow in cfm or cubic feet per second (cusecs) is found by multiplying the cross-sectional area of the stream by its velocity.

Mark off a section of the stream, at least 30 feet, where its course is reasonably straight and smooth. Choose a windless day to take the measurements. Place a float upstream of the first marking and time its passage over the known distance. A bottle, partially filled, and submerged to the 'shoulders,' makes an excellent float. Repeat the procedure and find the average time. Reduce this time by a correctional factor of 0.8 for a stream with a smooth bed and banks, and by 0.6 for a rock strewn hilly stream.

Next, the average depth of the river between its banks must be ascertained. This is done by taking a number of depth measurements across the bed of the river, at equal intervals, adding up all the readings and dividing the total by the number of measurements taken. The cross-sectional area, in square feet (or whatever), is then found by multiplying the depth by the width of the river.

Having ascertained the velocity and the cross-sectional area, the flow is found by multiplying the two.

Example: *A float on a river, with smooth bed and banks, takes 75 seconds to travel 50 feet. Thus the velocity is equal to:*

$$\frac{50 \times 0.8 \times 60}{75}$$

or 32 feet per minute. Ten cross-sectional readings were taken, giving a total figure of 14.5 feet. When divided by 10 this gives an average depth of 1.45 feet. The width of the river is 12 feet, so the cross-sectional area is 12 × 1.45 or 17.4 sq. ft. Therefore the flow is 32 × 17.4 or 557 cu ft/minute.

Measuring Head

The head (H) is the height the water falls from the headwater to the tailwater. The head exerts a pressure which can be turned into useful power. On high-head installations an indication of the head can sometimes be gained from detailed maps, but for a more correct measurement any of the following methods may be used:

1. Borrow or rent standard surveying equipment. The job re-

quires two people, one to hold the rod and the other to read through the transit. For the sake of accuracy it is wise to have someone on hand who has had experience with such equipment.

2. Use a hand level. A hand level is basically the looking glass part of a surveyor's transit. The tripod is replaced with a human body, of known height, so instead of using a staff one merely walks to the spot sighted through the level and then takes another sighting, and so on until the head is measured.

3. If you have time on your hands a good, though tedious method, is to tie a carpenter's level to a piece of straight board or light metal. Place the board horizontally (check with level) at the headwater, measure the vertical distance between the tip of the board and the ground, and keeping a record, repeat the process until the tailwater level is reached.

Having measured the gross head, we must now take into consideration the various losses which will result in a figure for the net head. There is always some loss of head on an overshot wheel installation as the wheel must run free of the tailrace. With impulse turbines, the Pelton and Turgo, there is loss of head for two reasons. First, there must be a gap between the nozzle jet and the tailwater. Second, head losses occur in pipelines due to friction. With PVC piping the friction is very low and even on installations with long pipelines I have rarely found the head loss to exceed 8 per cent. Any PVC pipe manufacturer ought to be able to give you a flow chart which will clearly show the head losses and also indicate the most appropriate pipe diameter to be used. Trying to cut down the capital costs by reducing the size of pipeline may ruin an otherwise excellent scheme. Steel, iron and concrete pipe all cause very high head losses, and as they cost more than PVC they are, at present, of little concern. Any pipeline with a large number of bends and undulations will have a bad effect on the flow and thus reduce the effective head considerably. The obvious choice for a water power site is where the highest head or fall is available in the straightest possible line, and within the shortest distance. Open chamber cross-flow, Francis and propellor turbines with draft tubes have little or no head losses. Those with pipelines suffer losses as above.

The net head (h) is the **actual** head or pressure available to drive the turbine or waterwheel when useless losses have been deducted from the gross head (H). Thus:

h = H — pipeline friction at full load and drop from turbine nozzle center line to tailwater level (impulse) or rise in tailwater level at full load (reaction).

Head Loss for Plastic Pipe

Flow cfm	3	6	12	18	24	30	36	42	48	54	60	66	72	78	84	90	120	150
Pipe Size						(Head loss per 1,000 ft. of pipe)												
2″	18	63	230															
2½″	6	21	75	161	274													
3″	2	9	30	64	110	166	234	312										
4″	½	2	7	15	26	40	56	74	95	118	144	172	201	230	268	305		
6″	0	¼	1	2	4	5	7	10	13	16	19	23	27	30	36	40	69	105
8″	0	0	¼	¼	½	1¼	1¾	2⅓	3	3¾	4½	5⅓	6⅓	7⅓	8½	9½	16	25

(Steel pipe in fair condition will have about twice the head loss shown above.)

With very low head turbine installations, say from 5 to 10 feet, the rise in tail water level must **not** be ignored. Even with a tailrace only a few hundred feet long, the tail water level at full load may easily rise 1 foot or more. Thus with an open type turbine in a pit, when the gross head is 6 feet from head water level to standing tail water level, the power from any type of turbine with 'h' reduced to 5 feet will be reduced by 24 per cent.

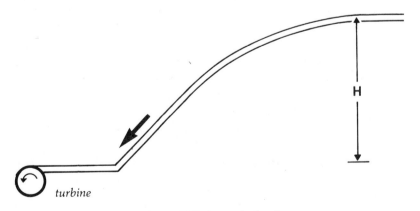

Fig. 2. "H" shows the head

How to Compute Theoretical and Net Power Output

Having calculated the flow (Q) and the net head (h) it is a simple matter to calculate the theoretical output (T) in kilowatts:

$$T.kW = \frac{Q \times h}{708} \quad \text{or} \quad = \frac{Q \times h}{11.8}$$

where Q = flow in cfm   or = flow in cusecs (cu ft/sec)
        h = head in feet   = head in feet
    708 = const. factor   11.8 = constant factor

The following equation is for those who have gone metric:

$$T.kW = \frac{Q \times h}{102}$$

where Q = flow in liters per second
        h = head in meters
    102 = constant factor

These equations show us the power available in flowing water if equipment with a 100% efficiency were available to tap it. However, as we haven't got that far yet, we must calculate according to available efficiencies. The maximum to be expected for a small water turbine is 80% efficiency; the figure drops to 65% for overshot water wheels.

Power transmission manufacturers are now claiming a 97% efficiency for their belt drives, so a turbine with one belt drive to the alternator will have a 97% transmission efficiency, with two belt drives a 94% efficiency and with three belts 91%. Gear-box manufacturers claim a 95%, or higher, efficiency. The efficiency of second-hand gear-boxes, bevel gears and motor vehicle back axles will vary slightly. A good alternator should have an efficiency of about 80% over a wide range of outputs. There is one on the market today with an 88% efficiency and another with a mere 71%.

From the above we can calculate that the overall efficiency for a water power installation using a turbine with a one belt drive and a good alternator will be 0.8 (turbine) × 0.97 (belt) × 0.8 (alternator) = 62%. The overall efficiency for an overshot water-wheel with a gear-box, 2 belt drives and an alternator will be 0.65 (water-wheel) × 0.95 (gear box) × 0.94 (2 belt drives) × 0.8 (alternator) =

46.5%. This indicates the two extremes in efficiency within which most installations fall. The high figure of 62% (which can be increased rarely in practice) is generally found on high-head, high-speed Pelton or Turgo impulse turbines. As the runner weight increases and its speed decreases on low-head installations, so the efficiency drops. Further information on efficiencies is given throughout the book in the description of each installation. It should be said that the efficiency of a turbine or water-wheel will fall if the flow or head for which it was designed decreases. If the shaft power to an alternator falls below 50% of its rated output its efficiency will begin to decrease. These efficiency figures for turbines and alternators are available from the manufacturers.

Example 1:   *The flow on a mountain stream is found to be 90 cubic feet a minute and the available net head is 180 feet. An impulse turbine with a single belt drive to the alternator is considered best, therefore the overall efficiency will be about 60%. The net output will be:*

$$\frac{90 \times 180 \times 0.6}{708} = 13.7\,\text{kW}$$

Example 2:   *A low-head reaction turbine with a gear-box and one belt drive is to be installed on a river with a 10 foot head and a flow of 50 cusecs (cfs). The net output will be:*

$$\frac{50 \times 10 \times 0.58}{11.8} = 24.5\,\text{kW}$$

WATER-WHEELS

During the 19th Century there were 23,000 water-wheels operating between Maine and Georgia alone, and how many thousands more must there have been throughout the world? Most of the wheels have gone, but in many cases the expensive civil engineering work remains, such as the leat or mill-race, wheel housing and tail-race. Many who own such sites are beginning to look at them with renewed interest, especially now that it is becoming economically viable and even desirable to install a new water-wheel or turbine. It is my opinion that, within the next 10 to 20 years, many of the old mills will be producing power again — this time in the form of electricity.

Water-wheels are simple, straightforward and graceful. It gives great pleasure (to me at least) just to sit and watch the wheel slowly turning as the water flows and splashes. The deep rumble of a revolving water-wheel is a far more pleasant sound than the high-pitched whine of a low-head turbine.

Types

The Overshot Wheel is the traditional, and probably the most efficient water-wheel. The water is supplied via a chute or flume to the buckets, which when full, fall, as a result of dead weight, thus turning the wheel. The supply of water to the wheel is regulated by a hand-operated sluice gate. The wheel itself must be clear of the tail-water. For example, on a 10 foot fall an overshot water-wheel should have a diameter of 9 ft. 3 in. or 9 ft. 6 in. Overshot wheels are suitable for heads from six to eight feet or more.

The efficiency of the overshot wheel is frequently grossly exaggerated. One source quotes 90%, whereas in fact it is usually between 60% and 65%. Rotational speed ranges from 6 rpm for large wheels up to 20 rpm for small ones. The main disadvantage of water-wheels is their slow speed. To take the speed from the shaft, up to the minimum of 1500 rpm required by the alternator, requires much gearing. It takes very heavy gearing to be capable of withstanding the high torque from the wheel. The initial step up is the most costly. Chain drive can have a limited life and the replacement cost is high. Most people use spur or helical gears, tractor gear boxes or truck back axles, giving a step up of between 4:1 and 12:1, and thereafter use multi V-belts through counter-shafts. A new gearbox, suitable for a 20 ft. wheel, could cost in the region of $6,400, and that would probably make an installation prohibitively expensive. The thing to do is to hunt around scrapyards and buy secondhand at a fraction of the new price. A good solid set of secondhand gears capable of long service is essential. There are a few water-wheel installations which use an external rack shroud gear on the outer rim of the wheel, driving a pinion.

The other disadvantage with water-wheels is their bulk, a lot of material and work goes into their construction. On the other hand maintenance is minimal and repair is simple. Low-head turbines are damaged by grit, are subject to cavitation (which destroys the runner) and their trash racks tend to get clogged up with leaves, etc. The water-wheel is not detrimentally affected by any of these ills. I have yet to see it proved, but it is a general opinion that the water-wheel will maintain its efficiency on fluctuating water flows better than Francis or fixed blade propeller turbines will. For the past 50 years or more, overshot water-wheels have been considered archaic by many associated with hydropower. This is mainly because of their bulk, slow speed and because they are not technologically complicated. I think the overshot does not deserve such scorn, and am delighted to see so many people installing their own water-wheels in the 1970's. It may be of interest to know that it required 3 kW of

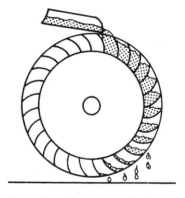

Fig. 3. Overshot water-wheel

energy to turn a standard mill-stone — but remember that the old mills did not grind night and day. I do not know of any way in which a water-wheel can be mechanically governed — nor do I want to. If a constant load is not available, an electronic load governor, or similar systems, should be used.

There are three main types of water-wheel which operate on heads too small for use with an overshot. Having a low output, they are of doubtful worth for the cost involved, especially when compared to the cost of a propeller, Francis or even a cross-flow turbine.

The Breast Wheel, is less efficient than the overshot, but involves the same amount of work to construct. Figure 4 shows a 'low' Breast wheel, though there are many examples of 'high' Breast wheels where the water enters the buckets at a higher angle.

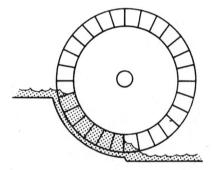

Fig. 4. Breast wheel

The Undershot Wheel, the most basic and primitive of all wheels, has a maximum actual efficiency of about 25%. If anything can be said in its favor it is that it can operate, or at least turn, on a one-foot head.

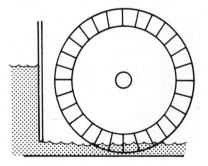

Fig. 5. Undershot wheel

The Poncelet Wheel, an improved undershot, is also the fore-runner of the cross-flow and Pelton. It depends not upon the dead weight of water in buckets, but upon the velocity of water forced through a narrow opening to strike the curved vanes. Efficiencies as high as 50% to 60% have been claimed for it. It is suitable for use on any head under six feet. Due to the fine clearances required, it is liable to damage from wood, stones, etc., carried in the water.

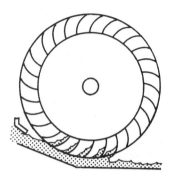

Fig. 6. Poncelet wheel

A Fiberglass Water-wheel Installation

Of the many water power sites I have visited in the preparation of this book, this installation was the most visually pleasing. The great 16 ft. diameter wheel and the intake flume are both deep green, and blend in beautifully with the trees and undergrowth of the surrounding landscape. The wheel turns at a stately eight revolutions per minute.

Photo 1 — Sixteen foot water-wheel

Photo 2 — The open channel takes water to the sluice gate where it enters the flume before filling the buckets on the wheel

The flume is placed further back on the wheel than is normal and this is with the idea of gaining additional power as the water strikes the blades of the buckets slightly before dead center. The shroud, extending from the flume over the top of the wheel, is to prevent water spray from the buckets. I am far from convinced that any gain is to be made from this method. Ideally the water should enter the buckets of an overshot wheel without shock, and the design should be such that the head above the point of entry to the wheel should be just sufficient to fill the buckets when the wheel is rotating at the correct speed. If the water comes as a powerful jet, it does not act like the jet on the buckets of an impulse turbine, but will entrain air, cause turbulence and blow the water out. Hold a wine glass under a fast running tap and see what happens. However, each to his own ideas, and it's good to see people experimenting.

Not only is this installation most attractive, but also the mechanical layout is simple and straightforward. The drive from the 6-inch wheel shaft is taken first to a tractor gear-box with a ratio of 20:1 and from there through two multi V-belt countershafts, which give the 1500 rpm required by the alternator. No governor is used or required on this installation. The sluice gate controls the rate of flow over the wheel which is matched to the constant load. The full electric output is constantly used.

There is an emergency trap door in the bottom of the flume, which when opened, discharges all the water before it strikes the wheel.

The exact efficiency of this installation has not yet been fully

Photo 3 — Tractor gear box (upper right) and alternator

ascertained. However, on a net head of 16 ft. and with a flow of 900 cfm it will produce at least 10 kW. Thus the overall efficiency is about 50%, broken down as follows: water-wheel 70%, gear up 90% and alternator 80%. The high efficiency of the wheel could be due to miscalculation, as the exact figures are not yet available, or it could be due to the lightweight design of the wheel, plus the fact that it has only 32 buckets which is considerably fewer than are found on old water-wheels. It would be possible to build an apron around the lower quarter of the wheel to prevent the water spilling out of the buckets until the last moment. Such a refinement could add to the efficiency.

The wheel, purchased from Westward Mouldings, Ltd., cost $2,960, the tractor gear-box $80, the V-belts, pulleys and second-hand generator cost $400, so the total cost was $3,440, to which must be added the cost of the following: sluice gate and intake flume, wheel support, discharge pit and powerhouse. The cost of these latter items to anybody else would depend on how much of the work they were prepared to do themselves and what materials they had at hand. All of this assumes that one has an old channel from which the water can be drawn, or else that you are one of the fortunate few who has a waterfall which could be tapped directly. Given good conditions and the provision of the right equipment, making a channel need not be too difficult a task for the healthy do-it-your-selfer, though I should hate to think what a civil engineering con-tractor's charges would be for the construction of the half-mile channel which serves this installation.

However, for a cost of $3,440 (£2,150) and assuming an average output of 5 kW this water-wheel provides electricity to the value of $1,752 per annum, assuming for the purpose of comparison a utility cost of 4¢ per kWh. As such, the payback period for the hard-ware is about two years. My one complaint is that the output from the 15 kW 3-phase alternator is not very efficiently used. The con-stant-load system means that lighting in the large estate is on both day and night. If an electronic load governor were installed then the power from the unnecessary daytime lighting could be diverted to more useful purposes — that is assuming there are other electrical requirements. Also with an electronic governor the voltage would be kept constant. At the moment it fluctuates when the water supply falls below normal. It should also be noted that the life of the tractor gear-box may be as short as one year.

Photo 4 — An old metal water-wheel with head tank giving pressure to the hose

An Old Water-wheel Installation

The photograph shows the 14-foot diameter wheel which has sheet steel buckets held in a cast iron frame, revolving on a 22-inch diameter cast iron shaft. The second-hand wheel has been running on this site for the past 30 years and looks fit to last at least another 30. It is not known when it was first built, but is at least 75 years old.

The 8 rpm drive is taken through a series of huge metal pulleys which are inside the house and bring the speed up to the 1500 rpm required by the alternator. The drive from the water-wheel is taken first through an 8-ft. diameter pit wheel to a 3-ft. bevel gear with a 4:1 ratio, then to a 5-ft. pulley, next a 7-ft. pulley and finally through a 6-inch pulley to the old alternator. I would like to have shown a photograph of these great spinning wheels, but as they extend over two rooms it would not be possible to do them justice.

All this equipment — and compared to the previous installation there is a vast amount of it — was picked up for next to nothing. The owner is, among other things, a professional wrecker. If it could be purchased new today the cost would be astronomic. There are 2,000 nuts and bolts in the water-wheel alone! The 3-ft. diameter bevel gear, a rarity, would probably cost new, in excess of $1,600.

Most of the old metal water-wheels have met with sorry ends at the hands of scrap merchants — yet there are some to be found. They can be a good buy, providing they suit the site and are in good condition. The odd thing about metal wheels is that they rust, not when the water is flowing over them but when it isn't.

The 9 kW output from this installation is used for domestic purposes in the house adjoining the wheel, and any excess power is used in the workshop. The steps on the embankment in the photograph lead to the reservoir, built in the days of grain milling. The reservoir is no more than 30 feet away from the house and is higher than the ground floor. Some years ago this caused a great deal of trouble and expense when the water began to break through the floor and form little springs — very pleasant, but not when in the living room. Considerable waterproofing had to be undertaken and the whole ground floor resurfaced.

Water-wheel Manufacturers

CAMPBELL WATER WHEEL COMPANY
420 SOUTH 42ND STREET,
PHILADELPHIA, PA., 19104
This company was founded in 1925 by John Blake Campbell, who at 87 is now the oldest manufacturer of water-wheels and the last of the old water-wheel craftsmen. In the past 60 years, he has been manufacturing, installing and renovating water-wheels. Overshot water-wheels are his specialty and these are manufactured to order. Turbines can be installed but are not manufactured by Campbell.

Westward Mouldings Limited,
Greenhill Works,
Delaware Road,
Gunnislake, Cornwall, England.
Westward Mouldings Ltd. manufacture a range of fiberglass water-wheels as follows:

Wheel diameter	8 feet	16 feet	20 feet
Number of buckets	16	32	40
Bucket capacity	3.6 cu. ft.	4.8 cu. ft.	7.2 cu. ft.
Max. advised rpm	15	10	6/8
Max. output (approx.)	3.6 kW	11 kW	25 kW
Price	$1,200	$3,200	$4,800
Price of home-assembly kit		$2,400	$3,600
(Suitable for importing)			

The maximum output in kilowatts shown above is based on a 65% water-wheel efficiency with the buckets filled to 70% of capacity, the figures exclude the increase in power which may result from the installation of an apron on the lower quarter of the wheel. The output relates to shaft power only, from which should be deducted generator and gear losses.

The calculations are based on the following equation:

$$\text{Output (kW)} = \frac{D \times B \times B.no \times rpm \times 0.65}{708}$$

where

D = wheel diameter from bucket centers
B = working bucket capacity (0.7 of total capacity, approximately)
$B.no$ = number of buckets
rpm = revolutions per minute
0.65 = efficiency factor
708 = kW conversion factor

Example: *A 16 ft. wheel with buckets filled to 70% capacity, 3.36 cu. ft., revolves at 8 rpm. Therefore its output is as follows:*

$$\text{Output} = \frac{14 \times 3.36 \times 32 \times 8 \times 0.65}{708}$$
$$= 11 \text{ kW shaft power.}$$

Build-It-Yourself Water-wheel Bibliography

"A DESIGN MANUAL FOR WATER WHEELS,"
W.G. OVENS. PUBLISHED BY VITA.
3706 RHODE ISLAND AVENUE, MT. RAINIER, MARYLAND 20822
This book concentrates on the design data and construction of overshot wheels. It is intended for mechanical drive in developing countries. It contains no information on gear-up to generators, but is otherwise excellent.

PAUL DILLOW,
2742 VICTORIA DRIVE,
ALPINE, CALIFORNIA 92001
Sells plans for alumninum wheels as described below for $10 each.

Wheel diameter	4 feet	6 feet	8 feet
Bucket width	1.5 feet	2 feet	2.5 feet
Bucket depth	0.5 feet	0.66 feet	0.66 feet
Maximum rpm	30	20	15
Required flow	180 cfm	330 cfm	420 cfm
Maximum shaft power	0.5 kW	1.4 kW	2.5 kW

NATIONAL CENTRE FOR ALTERNATIVE TECHNOLOGY,
MACHYNLLETH, POWYS, WALES.
They have recently installed a 10 ft. diameter wooden water-wheel. The plans are available for $1.00.

"WATER POWER. HYDRAULIC ENGINEERING 1899,"
REPRINTED IN ALTERNATIVE SOURCES OF ENERGY, NO 14,
ROUTE 2, BOX 90-A,
MILACA, MN 56353
$1.50
A good article, full of facts and figures for the construction of water-wheels.

"HANDBOOK OF HOMEMADE POWER,"
(See main bibliography). It includes plans for a 5 ft. diameter water-wheel with an output of between 0.3 and 0.7 kW.

"TREATISE ON MILLS AND MILLWORK,"
(See main bibliography).

Fig. 7. Pelton wheel with buckets on shaft in center

PELTON IMPULSE WHEEL

The Pelton Wheel was developed by Lester Pelton who patented his design in 1880. This wheel essentially depends on the impact of moving water upon curved buckets and not, like overshot water-wheels, upon buckets filled with water which move slowly as a result of dead weight. It has been called a developed undershot wheel. The Pelton is used wherever there exists a high head of water — at least 50 feet. The flow can be very small. For example, a mere 10 cfm on a 50 foot head will produce at least 350 watts.

It works as follows: Water is taken from a high head through a pressure pipeline with a narrow nozzle at the bottom. The water is forced under its own pressure through the nozzle to form a high velocity free jet which is directed onto the buckets of the Pelton wheel. As a result of the near perfect streaming in the buckets 80 to 90% of the energy of the jet is absorbed by the wheel. In the 10-20 kW range, efficiencies of 80 to 85% are normal, but on larger installations this figure can go up to 93%. One to four jets can be used per wheel, though I have never seen more than two on small installations. As with all other turbines, there is no "standard Pelton wheel," and whilst they can be bought off the shelf, there are many different sizes to suit differing conditions.

As the Pelton is comparatively small for its output, it can attain high speed and thus reduce or eliminate the need for gears. There are advantages in having a long pipeline leading to a turbine next to a house, especially in an isolated area. One is that the pipeline can be tapped and water drawn off under pressure for agricultural or domestic purposes. There may be in fact too much pressure, in which case a reducer must be used. The second advantage is that a similar arrangement could be made so that the pipeline serves as a fire extinguisher should the need arise. Moreover, one could build a swimming pool, Japanese garden or whatever, using the tail-water from the turbine. Water poses endless possibilities in the creation of aesthetically pleasing environments.

Unfortunately, long pipelines tend to be expensive. Thus the Pelton wheel is only useful in hilly or mountainous country. With plastic pipe the head loss due to friction is considerably less than it used to be with concrete and iron pipes. Even so, one must be cautious not to have a long low pipeline or too many bends in it. Another disadvantage with the Pelton lies in the source of power: some mountain streams tend to drop in flow in the summer. If this happens and a continuous supply of electricity is required, then batteries must be used to store power or a dam built to store water. Both options tend to be expensive.

Fortunately the requirement for domestic power in summer is usually very low. Even if there is no power available for part of the year, I would still encourage a close examination of the practicality of installing a turbine which would contribute to part of the yearly power requirement.

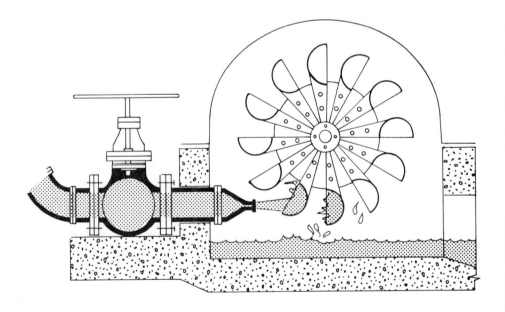

Fig. 8. The water jet streams out of the nozzle and strikes the curved buckets of the wheel

Pelton Installation No. 1.

Following is a description of a small Pelton Wheel installation used to power, through a battery/inverter system, a mountain homestead.

The photograph shows the simple siphon arrangement at the reservoir lake. The intake pipe extends to a depth of 6 ft. and has a protective wire mesh at its opening. The lake, a useful asset left over from past quarrying, contains 160,000 cu. ft., sufficient to power the system for five days. (It also provides cool but refreshing swims in summer.) Next to the intake pipe is a ball on a long arm, which registers the level of the lake on a panel at the bottom of the pipeline.

Four hundred feet of 4-inch diameter class B PVC piping, much of it laid on loose slag, carries the flow of 22 cfm to the turbine. The net head, after a loss of 4 ft. due to friction, is 95 ft. Output, at 50% overall efficiency, is 1.5 kW. This rather low efficiency is due to the fact that the second-hand equipment is not ideally suited to the site conditions. Seekers of second-hand equipment: good luck, but be careful. Take note of the warning on second-hand turbines on page 89.

Photo 5 — Syphon intake and water meter at the reservoir

Photo 6 — The old and over-rated alternator is in the center, Pelton wheel is under the wire mesh with part of the battery stock on the left

An earlier arrangement employing a spear valve that concentrated the jet of water in event of low flow, was removed in favor of full-speed operation or none at all. The battery bank takes up the slack. Should the reservoir run low, the gate valve shuts until the water level is high again.

The nozzle jet, moving at 53 mph, causes the 14-inch diameter bronze wheel to rotate at about 500 rpm. On the far side of the turbine, hidden by a wire mesh, is a belt drive giving a 3:1 gear up to the alternator. It took six men to lift the old and heavy 5 kW alternator into place. The alternator is both secondhand and over-rated for its purpose, but such a good solid machine with little work to do is likely to last a lifetime.

The efficiencies of both alternator and turbine on this site are 70%. The output, which rarely falls below 1.5 kW, is rectified from AC to 32 volts DC. Current is taken direct from the alternator as needed. Any excess current is fed into a battery bank for use later when the power requirement exceeds alternator output. Power used in the home and workshop is mainly DC, with 32-volt General

Electric incandescent lights used throughout, together with a host of old 32-volt equipment. An inverter is used for those items which require AC electricity at the standard 120 volts, 60 hertz, (or cycles).

The system outlined above is similar to that now available from either Independent Power Developers or Small Hydroelectric Systems and Equipment. Both manufacture mini-Pelton generating sets costing about $3,300 for the turbine DC generator, batteries and 3 kW inverter. An additional $1,000 or thereabouts would be needed for the pipeline, bringing the total cost to $4,300. If the continuous output of 1.5 kW is priced at 4¢ per kWh the payback period would be 8 years. The special stationary batteries have a life of about 15 years, so even if the cost of the whole plant is amortized over 15 years, that still leaves 7 years of free power.

Pelton Installation No. 2

This rather neat turbine set was installed in 1975. It is unique in that it is one of the first small turbines in the world to use an electronic load governor, as opposed to the more expensive and cumbersome mechanical governors. (More about that later.) The net head on this installation is 183 ft. and the flow 62 cfm, for an output of 8.75 kW, which gives an efficiency of 54%.

Photograph 7 shows the inlet (this installation has no reservoir). The water enters from the left and passes through the trash rack before leaving on the right. The rack, at an incline in the center, is cleared simply by pulling out the "plug" at the top lefthand corner, thus causing the water to flow backwards through the rack and so flush the debris down the drain. The plug, a section of PVC pipe with a handle, also serves as an overflow pipe. Moreover, this inlet arrangement acts as a settling tank for silt, which is essential on high-head installations wherever there is sand or grit suspended in the stream, as these materials can act as an abrasive force on the pipeline, nozzle and runner. Between the inlet and the turbine house runs 2,200 feet of 8-inch diameter class B PVC pipe. The full length of the pipeline is buried, a job that involved rock blasting. There is a 15 ft. head loss on the gross head of 198 feet.

In the powerhouse, photograph 8, the pipeline (encased in concrete) is reduced from 8 to 6 and then to 4 inches in diameter. In the same photograph can be seen the pressure gauge, main shutdown

Photo 7 — Inlet arrangement shows overflow pipe and trash rack, top of the pipe-line is half submerged on the right hand side

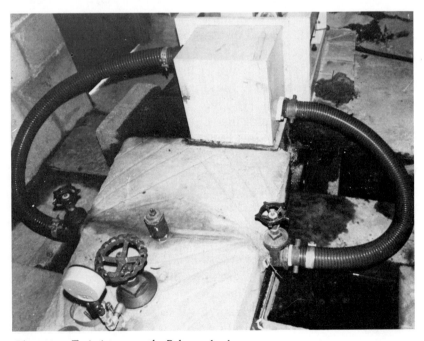

Photo 8 — Twin jets serve the Pelton wheel

valves, pressure release valve and the two flexible pipes leading to the Pelton wheel in the steel box. Each of these pipes has its own shutdown valve. Should there be a decrease in demand for power or a decrease in the supply of water, then one of the pipes is closed. This installation is also equipped with sleeves which can be inserted inside the nozzles, thus reducing the size of the jets or helping to form solid jets should the flow decrease. The Pelton wheel itself is 9.5 inches in diameter and is made of bronze. It revolves at 1300 rpm under a pressure of 83 lbs. per square inch and requires no gear-up to the alternator. As a result of using twin jets, the diameter of the wheel is half and its speed almost double what it would be using a single jet.

With this installation, there is a shaft leading from the Pelton wheel to an electro-magnetic clutch. This is an independently operated and essential safety device which, when necessary, disconnects the alternator from the shaft. Nuts and bolts between the alternator and its support are used to loosen or tighten the drive belt.

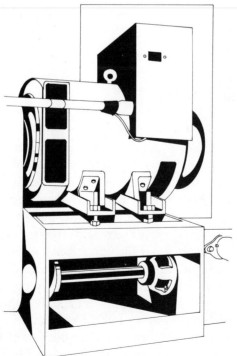

Fig. 9. Pelton wheel with attached alternator

The heavy duty alternator is rated at 12 kW. Governing is achieved by means of electronic load diversion. The main device used to effect this is no larger than a pocket calculator. Output from the alternator is used for cooking, space and water heating. The electronic load governor operates as follows: when the stove is turned on, sufficient current to meet the need is diverted from heating and when cooking is over the excess current is instantly diverted back to space or water heating. Thus all the power generated by the turbine is continuously used. Should the house get too hot then one of the nozzles is turned off manually or if the load in the house is rejected, then the clutch will disconnect the alternator. In addition, it is possible to have some form of resistance outside the house which will absorb any excess power: for example, greenhouse heating or a throwaway resistance.

In 1975 the hardware, including the electronic load governor and switchgear cost $2,640, the pipeline cost another $2,640 and was laid by the owner. A further $2,145 was spent on the hire of equipment, rock-blasting, building the inlet, powerhouse, etc. The total cost was $7,425 or $850 per installed kilowatt capacity. Assuming an average output of 5 kW this installation will pay for itself in just over four years.

Pelton Installation No. 3

Deney Smith built this system to power his remote homestead back in 1951, and the trouble-free turbine continues to fulfill its function. The source of power flows from a clear mountain spring, sited some 420 feet above the power house. The flow is taken, first to a silt trap, and from there it enters a six-inch diameter steel pipeline, 1,800 feet long. Much of the pipeline, laid on rough ground, is now overgrown with moss, fern and briars. Back in 1950 the steel pipe cost 25¢ a foot. Four-inch diameter pipe was laid initially, for the first few hundred feet, but that was abandoned when the owner realized that the head loss on such a pipe due to friction would render the whole installation futile. The importance of this lesson cannot be overstressed, if the pipe diameter is too small, friction losses increase and the turbine output is radically lowered.

Recently two small leaks were found in the pipeline. In relation to this, it should be noted that if a shut-off valve is installed at the top of a line, then a ventilation valve should be installed near the shut-off point, to dry out the pipe for repair purposes. This is to prevent the pipe from collapsing in on itself.

The runner is an 18-inch diameter Pelton wheel, made of steel casting, weighing 85 pounds and having 15 buckets. A 1¾-inch stressproof steel shaft carries the wheel. The seals are of the water slinger type and seem to last forever. One end of the shaft is connected to a Woodward UG 8 governor which maintains constant voltage and frequency against varying loads by diverting the water on and off the Pelton wheel. Connected to the other end of the shaft is a V belt and pulley which drives the old General Electric alternator at 1,200 rpm.

After about 25 years the governor was replaced, not a very difficult or expensive job. The V belt was replaced once. The original still hangs on the wall just in case the new one breaks. The only maintenance on the alternator, apart from greasing, is replacing carbon brushes for the field voltage. After a quarter of a century of constant use, day and night, all the equipment remains in excellent condition. Remarkable though it may seem, the bearings still do not need replacement. The Pelton wheel and the nozzle are in perfect condition, undamaged by the clear spring water. The importance of using a silt trap to prevent damage from grit can be seen from this.

Photo 9 — A 15 kW Pelton wheel with Woodward governor

Output from the alternator, which reaches a maximum of 17 kW in winter, supplies all the electricity needed by the original house. A second house, recently built nearby, is also powered from the same turbine, and excess electricity is used to heat the greenhouse. An L.P. gas back-up heat and cooking system is installed but rarely used. The cost of the turbine built by Smith back in 1951 was $3,500, and that excluded the governor and alternator. Today an 18-inch Pelton turbine from Small Hydroelectric Systems and Equipment, in Washington state, (see Manufacturers), which specializes in such Pelton equipment, would cost less than it did in 1951, though admittedly not made from steel castings. The governor would cost approximately $500.

Having powered the turbine, the water then flows into two ponds, both stocked with fish, before returning to its original course.

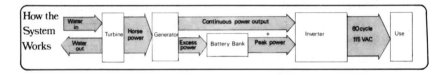

Fig. 10. The Pelton wheel system

Pelton Manufacturers

INDEPENDENT POWER DEVELOPERS, INC.,
BOX 1467
NOXON, MONTANA 59853
(406) 847-2315
IPD was founded by Bill Delph, a master electrician who first worked for a large hydroelectric power company in Washington State. IPD now produce two turbines, one a Pelton and the other a low-head propeller type (see p. 61 under Francis and Propeller Turbine Manufacturers). The tiny 4½-inch epoxy-coated silicon aluminum alloy wheel is directly connected to the generator.

Head	Flow	Output
46 foot	3 cfm	125 watts
103 foot	4 cfm	380 watts
160 foot	5 cfm	750 watts
290 foot	7 cfm	1900 watts

The heavy duty Motorola generator produces a maximum of 70 amps at 32 volts DC. With an output of 2 kW, or less, it is sufficient to satisfy the electric needs of a small house, provided some other fuel is used for heating. The turbine/generator set sells for $985. Batteries suitable for use with this system cost about $1,000 and a 3 kW inverter sells for $1,400. Add about $1,000 for a good pipeline and that brings the total to $4,400. Two or more units can be coupled together to provide additional power, and cost per installed kilowatt tends to decrease when multiple units are used. Independent Power Developers were recently awarded $20,000 by Montana Dept. of Natural Resources to build three small-scale hydroelectric plants. Send $2.00 for their water power catalogue.

SMALL HYDROELECTRIC SYSTEMS & EQUIPMENT
P.O. BOX 124
CUSTER, WASHINGTON 98240
(206) 366-7203
SHSE manufacture a complete range of Pelton wheels. They also sell the basic drawings and parts for their home construction.

Pelton wheel castings	Aluminum Alloy	Bronze
4½-inch casting	$125	$250
9-inch casting	$150	$300
18-inch casting mounted on a 2-inch shaft.	$450	Price on request

The prices shown above do not include the housing, bearings or nozzle.

Ready-to-run turbines in housing	Aluminum Alloy	Bronze
4½-inch wheel, "Water-Lite" with 4 nozzles and alternator mounting plate	$500	$650

9-inch wheel with 1-inch shaft and bearings	$750	—
9-inch wheel with $1\frac{3}{16}$-inch shaft and bearings	—	$1,250
Two 9-inch wheels with $1\frac{3}{16}$-inch shaft and bearings	$1,000	—
Two 9-inch wheels with $1\frac{7}{16}$-inch shaft and bearings	—	$1,500
Complete 15 kW Pelton wheel hydro-electric plant with Lima brushless alternator and Woodward governor (a diesel plant of the same output could cost more)		from $6,000

SHSE also stock a comprehensive range of DC and AC generators, batteries and inverters suitable for use with the miniature "Water-Lite" unit, designed to produce up to two or three kilowatt. The beauty of the inexpensive Water-Lite unit is that it has four jets, one or all of which can operate at any given time. In effect, this means that the versatile turbine will produce power on low or high heads, and the number of jets can be varied to suit seasonal changes in water flow.

SHSE manufacture a range of Pelton wheels capable of producing from a few hundred watts to a few hundred kilowatts. Send $2.00 for information package.

Photo 10 — Bill Kitching of SHSE with complete 15 kW plant and 9 and 18 inch Pelton castings

ALASKA WIND & WATER POWER
P.O. BOX G,
CHUGIAK, ALASKA 99567
(907) 688-2896
Francis Soltis of Alaska Wind & Water Power worked closely with Bill Kitching of SHSE in the development of their range of Pelton wheel systems. As an engineer, he has been involved in small water power for over seven years now. His Alaskan company markets and installs the complete range of tubines manufactured at the SHSE foundry. Send $2.00 for details.

PUMPS, PIPE AND POWER
KINGSTON VILLAGE
AUSTIN, NEV. 89310
(702) 964-2483
PPP supply Pelton wheels with outputs from 15 kW upwards. They use Woodward governors for speed control. PPP practice what they preach and generate their own power with a Pelton wheel operating under 450 feet of head.

ELEKTRO G.M.B.H.,
ST. GALLERSTRASSE 27,
WINTERTHUR, SWITZERLAND.
Elektro have been manufacturing turbines and wind generators for the past 30 years. Apart from their range of Peltons they also manufacture a small Francis turbine, which is shown on page 56. Their four Pelton wheels range in cost from $1,100 for 300 watts to $12,000 for up to 24 kW, which is more expensive than US-made turbines. Elektro has a simple way of improving the output of small Pelton wheels during times when the flow is low. It is to build a small reservoir capable of containing the normal volume of water that would flow in 15 to 30 minutes — i.e., for a flow of 3 cfm the reservoir would contain 45 to 90 cubic feet. When water supply is very low then the Pelton wheel efficiency drops considerably. To prevent this, a ball-valve closes the pipeline allowing the reservoir to fill. When it re-opens the Pelton drives at full efficiency until the reservoir is emptied and then the cycle begins again. This means that instead of a continuous output of, say, 25 watts (600 watt hours per day), output can be increased to eight hours at 300 watts (2400 watt hours per day).

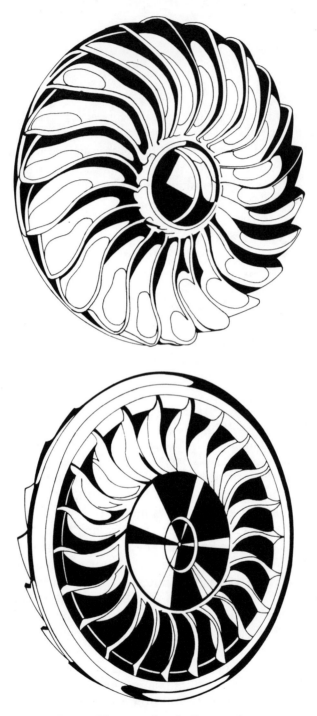

Fig. 11. The two sides of a Turgo turbine

TURGO IMPULSE WHEEL

The Turgo runner is basically an improvement on the Pelton. It was designed in 1920 by Eric Crewdson, then the managing director of Gilbert Gilkes and Gordon Ltd.

With the Turgo, the jet is set at an angle to the face of the runner, strikes the "buckets" at the front, and discharges at the opposite side. The basic difference between the Turgo and the Pelton will be clear from Figure 12.

Any impulse wheel achieves its maximum efficiency when the velocity of the runner at the center line of the jet is half the jet velocity. Hence for maximum speed of rotation the diameter of the runner should be as small as possible, and so the ratio of the runner diameter to the jet diameter is critical. The Pelton has a minimum runner to jet ratio of 9:1.

Crewdson set out to design a runner which would operate on a reduced ratio and thus increase the speed. The successful outcome of his endeavors was the Turgo, with a minimum runner-to-jet ratio of 4:1. In effect the Turgo runs at twice the speed and is only half the diameter of the Pelton. Therefore the necessity for gears to the generator is greatly reduced, as is the manufacturing cost of the runner itself. It can be served by one or two jets, has an efficiency of over 80% with a high part-gate efficiency and is suitable for use on heads of 40 feet or more. The Turgo is in use all over the world and has established a good reputation for trouble-free operation.

Example: *Assuming a net head of 100 ft. the nozzle velocity (v) will be the square root of:*
$2 \times 32 \times 100$ ft. sec. $= 80$ ft. sec. $= 4800$ ft. min. (Water, streaming out of a nozzle at the end of a pressure pipeline, has the same velocity as if it had fallen

from the height of the water level. The velocity rises in proportion to the square root of the height.) Runner rim speed = 0.5 × v = 0.5 × 4800 = 2400 ft. min.

Pelton	**Turgo**

Runner diameter 16 inches
Circumference =

$$\frac{3.14 \times 16}{12}$$

= 4.18 ft.

rpm = *rim speed / circumference*

= $\frac{2400}{4.18}$

= 574 rpm

Runner diameter 8 inches
Circumference =

$$\frac{3.14 \times 8}{12}$$

= 2.09 ft.

= $\frac{2400}{2.09}$

= 1148 rpm

Pelton

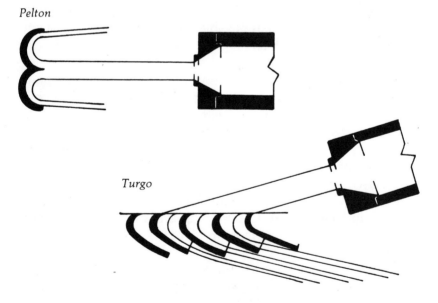

Turgo

Fig. 12. Turgo and Pelton turbines contrasted. The jet on the Turgo strikes three buckets continuously, whereas on the Pelton it strikes only one. A similar speed increasing effect can be had on the Pelton by adding another jet or two.

Turgo Installation

It was in December, 1962 when the owner of this installation first considered in detail the practicality of generating electricity from the stream which ran through his land. Shortly afterwards, having decided to go ahead, he surveyed the site, drew plans, submitted them, and in May 1963 they were passed by the local council. Work began the following month, with the owner working weekends and his two sons full time on the job.

First a small dam was built to accommodate two trash racks and the pipe inlet. Next, 400 yards of 15-inch diameter concrete pipe was laid along the bed of an old mill race which used to bring water to a fifty-foot water-wheel attached to a woolen mill. Nothing now remains of either the wheel or the mill, but the original mill race still survived for use after some cleaning out. The clay bed of the race was dug out, the pipes laid and then covered over. This was not an easy job since each pipe section weighed 640 lbs. and the only help available was a hand-operated winch. At one point the mill race had subsided and the only way to maintain the gradient required to carry the water to the head tank was to tunnel through solid rock for about 20 feet. This was done with hammer and chisel.

The head tank is 12 feet high, which is equal to the total head from the inlet to the tank. Its primary purpose is to facilitate the transition from the 15-in. low-pressure pipe to the 10-in. high-pressure pipe and stabilize the flow of water. It has a secondary purpose of preventing dangerous pressure surges occurring in the pipeline should anyone shut the spear or main valves. The head tank could be compared to the safety valve on a pressure cooker.

From the top of the tank to the powerhouse is a near vertical drop of 97 feet, down which was laid 10 in.-diameter metal pipe. It took just over three months to complete all the pipework. In November the owner and one son built the turbine house, using old stone from the ruined mill. The turbine arrived in December and first produced power in January, 1964.

Water is taken at a maximum of 210 cfm which turns the 13½-in. mean diameter Turgo impulse wheel at 500 rpm. This is speeded up through a 3:1 gear, using belt drive, to generate between 16 and 18 kW maximum at 240 volts A.C. Even during a summer drought the flow never falls below 70 cfm, which produces 6 kW. Calculated on the head of 97 feet, the overall efficiency of this in-

Photo 11 — Turgo turbine with alternator and flow governor

stallation is 64%. Output from the turbine is used for domestic purposes in the owner's home. When first installed the output was 19 kW maximum, but due to age and deposits in the pipeline this has decreased slightly to about 18 kW. Properly maintained it is probable that this installation will last at least another 50 years. The wheel bearings need occasional lubrication and the bearings themselves require changing once every seven years. Life of the alternator may well be 25 years and could even exceed this. The flow diverter governor may need occasional adjustment. During autumn and winter the trash racks need weekly cleaning. However this need only be done once every few weeks throughout the rest of the year. The actual cleaning, using a rake, takes five minutes at the most.

The total cost of the installation in 1963 was $4,800 with the turbine, governor and alternator costing $3,200 and the pipeline $1,600.

At today's prices the governor and turbine would cost $11,200. A similar alternator, from Lima Electric Co. (see Alternators and Generators), would cost about $960. The low pressure pipeline could be served by galvanized steel spiral wound tube of the type used for French drains and ventilation purposes. This would be considerably cheaper than PVC or concrete pipe. It would cost $2,554 for 1200 ft. from Metal Sections Ltd. (U.K.). The rest of the pipeline serving the 97 ft. head would have to be class B PVC at a cost of about $420. Add, say, $800 for the intake works, trash rack, turbine house, electric cable etc., and this gives a total cost in the region of $16,000 for the installation, or approximately $880 per installed kilowatt. The average output is 10 kW which when valued at the grid price of 4¢ per kWh is worth $2,800 per annum. Thus this installation at today's prices would pay for itself in under six years. The original $4,800 installation has paid for itself many times within its thirteen years of operation.

It seems very expensive to invest $16,000 in a water turbine installation, but those who spend such money now will look back at some future date and congratulate themselves on a sound investment. It is likely that this turbine will still be working long after the North Sea oil is exhausted. As for the initial hard work involved, and there was no lack of it in this case, the owner-builder says he was never so hale and hearty as when in the thick of it.

The Hydec Turbine Set

This is a recent and welcome addition to the group of turbines available today. Within its range on medium to high heads it is competitively priced against the Francis and Pelton turbines. Its design and construction is simple, as is the installation and maintenance of the unit. The efficiency of the Hydec is about 80%, which is average for most small turbines. The runner is the Gilkes Turgo Impulse Wheel. The wheel and its casing are both made of cast iron. The shaft is steel and the governor is an oil spring-loaded type which operates a stainless steel jet deflector. The inlet valve is a manually operated butterfly valve. Its output range is from 5 kW under heads as low as 40 ft. up to 150 kW under a head of 350 ft.

Cost

Mean Diameter of Runner	Single Jet	Twin Jet
7.5 in.	$7,840	$9,300
10.5 in.	$9,760	$11,500
13.0 in.	$11,200	$13,000
16.5 in.	$12,500	$14,700

These prices include the wheel, complete with casing, governor, inlet valve and inlet pipe. Suitable governors and control panels can be supplied at additional cost. But, to quote an example, a 16.5-in. twin jet unit developing 25 kW on a net head of 40 ft. was recently sold, complete with generator and switch panel, for $19,200.

The Hydec is available from:

Gilbert Gilkes and Gordon Ltd.,
Kendal, Westmorland,
England.

Gilkes was founded in 1856 and they are the oldest manufacturers of water turbines in the world. They have kept a record of every turbine manufactured by them and can tell just what head and flow each was designed to operate under. In fact, they still have the original design for their turbine No. 1, a 4 kW Thomson Vortex, built in 1856, and which operated for over a century. May the same be said a hundred years hence of their new Hydec range.

They also manufacture a complete range of Francis and Pelton turbines to order. Unfortunately they tend to be very expensive per installed kilowatt as compared to the Hydec. They are also developing an electronic load governor.

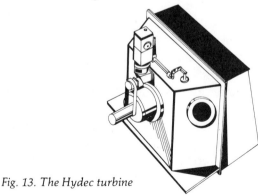

Fig. 13. The Hydec turbine

Fig. 14. The cross-flow turbine is split to facilitate flow-governing of incoming water

THE CROSS-FLOW TURBINE

This turbine is suitable for use on a wide range of heads (3 to 600 ft.) and flow (60 to 15,000 cfm). In its range it is similar to the Francis turbine but it is far simpler to construct and repair, and has a higher efficiency on part flow. The cross-flow is a radial impulse type turbine. Its characteristic speed places it between the Pelton and the Francis turbines. The jet of water, which is given a rectangular cross-section at the turbine inlet, flows through a ring of blades on the barrel-shaped runner, first from outside to inside and then (after crossing the interior) from inside to outside again. In effect it is an advanced Poncelet water-wheel with an efficiency of 80% on small plants and 84 to 88% on medium-to-large units.

In 1903 an Australian, A.G.M. Mitchell, patented a cross-flow turbine. About 20 years later Ossberger (see Manufacturers) began to manufacture the turbine, and as a result of considerable research and development they now produce an efficient type of runner and well-designed turbine sets. Professor Donat Banki, in Budapest, also developed a cross-flow turbine and published the details in 1917. Three years later Ganz Mavag (Budapest) went into the production of cross-flow turbines, but regrettably production ceased in 1946.

One definite advantage that the cross-flow has over both the Francis and propeller turbines is that it is not subject to cavitation. Leaves and other 'soft' trash also pass through the blades of the cross-flow without causing damage. The efficiency of the cross-flow is reduced if the runner is flooded, however, and therefore it must be placed above the tail-water level. A draft-tube can be placed between the turbine and tail-water, thus creating a suction head which compensates for the loss of head.

The cross-flow is the simplest turbine to manufacture, and those with good workshop facilities should have no difficulty in making their own. The only disadvantages that I can think of are that it does not have as high an efficiency at full flow as the Francis, nor is its draft-tube effect as good as with the Francis. One odd thing is that very few of the standard books on hydraulic engineering mention the cross-flow at all.

Ossberger Cross-Flow Installation

Out on the west coast of Ireland, John Wood has installed his own hydro-electric plant, which for the past 19 years has supplied both heat and outside lighting for his hotel, "The Falls" at Enninstymon, Co. Clare. At first an old Francis turbine was used to drive a DC generator, but that was replaced in 1967 by a more efficient Ossberger turbine driving an AEG 40 kW, three phase alternator. The Ossberger cross-flow turbine was specifically designed for this site, whereas the old second-hand Francis was not.

The net head of 15 ft., with a minimum flow of 3,000 cfm, gives an output of 35 kW, and thus the efficiency of the whole installation is 55%. Photograph 12 shows the turbine house on the left and next to it the Denil-type fish pass built by the Ministry of Fisheries and Agriculture to make life for the salmon easier. The river is tidal to the foot of the falls. When turbines are located on the river, as this one is, they have to be built very securely so as to withstand the great force of flood waters. During winter the rocks to the right in the photograph are completely covered in rushing water. Part of the solid rock waterfall had to be blasted away so that firm foundations could be laid.

The installation has no reservoir. It depends upon a spate river which, though generally full in the winter, can get very low in summer. Therefore, the output ranges from a maximum 35 kW in winter down to 10 kW in summer. There is usually a period of 4 to 6 weeks in the middle of summer when the river runs dry. This period is used to overhaul the plant. Fortunately the requirement for heat and light at that time in summer is negligible.

The runner, inlet and casing were purchased in 1967 from Ossberger for $2,700. With remarkable luck, Mr. Wood picked up, second-hand but unused, just the right size set of gears, the governor

Photo 12 — *Cross-flow installation on a waterfall; grid at end of tail-race diverts salmon safely up river*

and flywheel, plus the alternator, all for a mere $150. The same equipment purchased from Ossberger today would cost $28,000. It is my opinion that the horribly complicated and expensive mechanical governor (with flywheel) which Ossberger supplies is quite unnecessary — it would be incomparably cheaper to use an electronic load governor instead. Sad to say, there is no sign of Ossberger marketing such equipment.

The installation took three months to complete and cost a total of $5,600. The plant in the turbine house is mounted on a large truck frame, slideable to facilitate belt tensioning. As a result of the conversion from DC to AC the switch boards had to be updated and new line poles installed to carry four cables. Two men raised five standard electric poles and secured them with tie wire anchors within 5½ hours! The two poles on the rocks were a different kettle of fish. The isolating switches, bus bar boxes, etc. for the switchboard were all second-hand from a defunct factory. Phase meters are at the hotel end, with only the voltmeter in the turbine house.

Photo 13 — Some of the equipment in the turbine room showing mechanical governor, flywheel, switch box and alternator; both the turbine and gear box are out of the picture on the right

All the work was done by the owner, his son and hotel gardeners, with some valuable help in the last stages from a ship's engineer on holiday. The local tombstone carver lent his lifting tackle for moving the hardware into place.

As said before, a plant similar to that which Mr. Wood purchased for less than $3,000 in 1967 would now cost $28,000, add to that, say, $8,000 for the civil engineering (which in 1967 cost $2,900) therefore the same installation today would cost $36,000. The average output from this turbine is 15 kW which, at 4¢ per unit, is worth $5,260 per annum. If it was installed today the payback period would be 7 years. Mr. Wood's installation has paid for itself many times over in its eleven years of operation.

This cost could be substantially reduced for those with an engineering background or a mechanical flair. A basic turbine similar to the one used by Mr. Wood, can be had from Ossberger complete with base frame, rectangular reducer and inlet pipe, draft tube and some coupling for approximately $8,800. Providing there is at least one over-speed safety device, a suitable 3 phase, 40 kW alternator could be purchased for about $1,000.

From there on it is the luck of the draw, plus mechanical ingenuity. A good second-hand gear box, which would take the 200 rpm shaft speed up to or near the alternator speed, can sometimes be found quite cheaply. But if that is not possible or too expensive, then a series of belt drives could be used instead.

The hydro-mechanical governor and the flywheel would be unnecessary. If a constant load system would be impractical then $1,600 plus should be set aside for an electronic load governor, switch board and safety devices. As such, the total cost for the hardware could be reduced to below $12,000, representing a saving of at least $16,000. Even so, I would only recommend this way to those who have had mechanical or engineering experience and are prepared to work hard at it.

One of the interesting aspects of this installation is the homemade automatic trash cleaner. The cleaner itself is made from car tires cut to give a tooth edge and held horizontally with metal strips. The cleaner is connected by a pulley to a 50-gallon tank. The tank is filled by gravity and when full it falls, thus drawing the cleaner up the trash rack and flipping the debris into a special trough. The 50-gallon tank empties by syphonic action and is then drawn back upwards by the weight of the cleaner — and so the action repeats itself every 14 minutes. The turbine is placed downstream from a town, which results in a fine collection of trash — everything from dead cats to food cans. It is important to keep the rack clear, as otherwise the flow of water will be impeded and the power output reduced. This installation also has a coarse trash rack which prevents damage from logs and other heavy items which tend to get swept along in a spate.

Apart from the mechanical governor, I like this run-of-the-river installation a lot. There are advantages inherent in the site: as the turbine fits snugly into the waterfall no expensive pipeline or headrace is required; moreover there is no need for a dam or reservoir.

A Small Owner-Built Cross-Flow Turbine

The photograph shows a cross-flow turbine being built from the Oregon University paper (see Bibliography). The actual blades of the runner were cut with a band-saw from 4-in.-diameter mild steel pipe, four blades to each length of pipe. The two 12-in. diameter end discs were flame cut from ¼-in. mild sheet steel. The slots in the discs for the blades and the central holes for the 1-in. steel shaft were cut with a milling machine. The blades were brazed into place. The rough edges on the outside of the rim were then ground smooth. The clamps on the shaft, which fit inside the bearings, are to stop the shaft from floating. The manually adjusted water intake will be made from sheet metal or resin and fiberglass.

The speed of the runner is about 300 rpm, which is taken up to 1800 rpm to the generator through a V-belt and pulley with a ratio of 6:1. The turbine operates on a 15 ft. head with a flow of 60 cfm, giving an output of 0.75 kW at an overall efficiency of 59%. Output from the slightly overrated generator is currently on a constant DC water heating load, and so no governor is required.

Some time in the future a battery system will be used to store the daily output of 18 kWh for peak domestic needs, such as cooking, lighting, etc. Meanwhile the owner is hunting around for old and new DC appliances, the idea being to try and avoid the expense of a DC-to-AC inverter. Plans are also afoot to install solar panels for summer water heating and a wood burning boiler for winter space and water heating.

Photo 14 — Cross-flow runner with shaft, pulley and bearings

Cross-Flow Turbine Manufacturers

BELL HYDROELECTRIC
3 LEATHERSTOCKING STREET,
COOPERSTOWN, N.Y. 13326
(607) 547-5260

Bell Hydroelectric manufactures a complete range of cross-flow turbines, a very exciting addition to the now complete range of water turbines manufactured in America. Truth is that Ossbergers are expensive, and importation from Germany takes time and adds expense.

Bell Hydroelectric manufactures its turbines individually and to order, hence prices are not readily available. But as an indication of cost an installation similar to the previous Ossberger installation would be around $20,000. Bell cross-flow turbines may be used

on a wide variety of heads from 5 to 350 foot, with outputs from
2 kW upwards. It is generally found that the cross-flow is most
cost-effective when used on the medium-head range, i.e. from 15
to 50 or 100 ft. Propeller turbines are cheaper on very low heads
and Pelton wheels on the very high heads.

James Bell, the hydroelectric engineer who started the company,
will be glad to send details of his turbines ($2.00) and at the same
time will give a cost estimate for any particular site if given details
of head, flow and power required.

OSSBERGER-TURBINENFABRIK,
D — 8832 WEISSENBURG 1. BAY,
POSTFACH 425,
WEST GERMANY.

There are over 7,000 Ossberger turbines installed all over the world,
with outputs ranging from 1 kW to 1000 kW. As each installation
is different, they do not issue price lists, but they are quick to quote
costs when given details of head and flow. The following will give
a general indication of their prices:

Output	Head	Cost per kW (installed)
25 kW	16 ft.	DM 2100 ($840)
25 kW	160 ft.	DM 1500 ($600)
250 kW	16 ft.	DM 840 ($340)
250 kW	160 ft.	DM 520 ($200)

The above prices include the turbine, frame, connecting piping,
draft-tube, gearing, automatic mechanical governor, flywheel, three-
phase generator, and switchboard. Not included are screens and
pipelines. Further prices are quoted in "Ossberger Cross-Flow Installation."

Ossberger have increased the efficiency of their runner on part-
gate (part-flow) by dividing it into 2 or 3 sections or cells. For example, should the flow decrease to one third of capacity, then only one
out of three cells is driven, which is more efficient than trying to
drive all three.

Ossberger also manufacture three standard turbine sets — the
Universal A, Universal B and the Hydro-Light. The Universals
operate within the following ranges:

Universal A

Head in Feet	13	20	26	33
Flow in cfm	104	127	148	170
Rpm	380	470	550	615
Shaft power in kW	1.5	2.8	4.5	6.3

Universal B

Head in Feet	33	66	99
Flow in cfm	47	66	80
Rpm	615	870	1070
Shaft power in kW	1.7	5	9

Both the models cost around DM 9600 ($4,400) excluding generator and gearing.

The third standard turbine set is the Hydro-Light, which has a manual governor and comes as a complete unit with generator for a cost of approximately DM 20,000 ($8,000) (excluding pipeline). Suitable for use on heads of 12 to 29 feet with a flow rate of 60 to 170 cfm, its maximum output is 5 kW. The manual governor can be controlled by long distance cable drive, though such governing would be unnecessary with an electronic load governor.

Finally, it has to be said that Ossberger has a reputation for good engineering.

BALAJU YANTRA SHALA (P) LTD.,
BALAJU,
KATMANDU, NEPAL.
BYS manufacture a range of nine small cross-flow turbines. In Nepal these turbines are used for mechanical purposes. If you are interested, write to them, enclosing a dollar or two and ask for details. Do not expect a quick answer though, since they took nine months to reply to my letter.

Build-It-Yourself Cross-Flow Bibliography

"THE BANKI WATER TURBINE,"
BULLETIN SERIES NO. 25.1949,
SCHOOL OF ENGINEERING,
OREGON STATE UNIVERSITY,
219 COVELL HALL,
CORVALLIS, OREGON 97331
Contains a translation from the German paper by Donat Banki
'Niue Wasser-turbine' (Budapest 1917) together with tests on a cross-
flow runner which at low efficiency (65%) produced 2 kW at 280
rpm using 133 cfm on a 16 ft. head. The theory of the Banki is well
covered, and instructions for the construction of the above runner
are included. Good value for money at fifty cents.

"LOW-COST DEVELOPMENT OF SMALL WATER-POWER SITES,"
H.W. HAMM.
Available from V.I.T.A., 3706 Rhode Island Avenue, Mt. Rainier,
Maryland 20822. It includes details on how to build a 12 in. cross-
flow runner.

"B.Y.S. CROSS-FLOW TURBINES."
A joint development project of N.I.D.C. and S.A.T.A. (1974) by
Clemens Adam.
This paper was produced to show a 'machine shop' in Nepal how
to build cross-flow turbines in the 10, 20 and 40 kW range. Highly
technical and useful to those who want to start production of such
turbines. Only 50 copies of this 51-page paper were printed, but
other copies may be available from: Hel Vetas, A Sylstrasse 41,
Zurich, Switzerland or Clemens Adam, 9320 Arbon, Switzerland.
I suggest that $10 should be remitted with each request for a copy.

"THE CROSS-FLOW TURBINE." *WATER POWER*
(*Now INTERNATIONAL WATER POWER AND DAM CONSTRUCTION*).
JANUARY, 1960 (see General Bibliography).
It contains a technical description of the Cross-Flow from the
Mitchell and Ossberger point of view. Reprints are available from
Ossberger.

Fig. 15. Francis turbine in open chamber with guide vanes and draft channel

THE FRANCIS TURBINE

The Francis turbine was designed and developed by James B. Francis, an Englishman who settled in the United States, and where he did most of his work. It is suitable for use on a wide range of heads, from 4 ft. upwards. Due to its high efficiency on full-flow the Francis quickly became popular, and today it is frequently found in use as the prime mover in megawatt installations.

On low-head large-flow installations the turbine is mounted in an open chamber. The water is directed onto the blades of the runner by the guide vanes on the periphery. Where a mechanical governor is used it is connected to the guide vanes, which in turn regulate the flow of water to the runner. The water, having given up most of its energy to the runner, falls into a draft-tube. The tube enables the fall below the turbine to be used as effectively as if it were above the turbine. Hence the turbine can be placed higher above the tailwater level than otherwise would be possible. However, it is not advisable to have a suction head of more than 20 ft. The turbine shown opposite has a runner diameter of 1 ft. and is a copy of one which produces 3.2 kW at 270 rpm on a net head of 5 ft. with a flow of 570 cfm and an efficiency of 80%.

Photograph 15 shows the spiral casing with guide vanes, as used on high-head Francis installations. Photograph 16 shows a 35 kW installation on a 46 ft. head. On the left is the spiral casing and in the center is the complicated oil pressure governor with a 670 lb. flywheel. The governor adjusts the guide vanes to a fine degree so that the generator output agrees with the load. All that complicated and expensive hardware could now be replaced by a small 'black box' containing an electronic load governor which would divert the excess load to some other resistance. The runner on this installation has a speed of 1,200 rpm.

Photo 15 — Spiral casing with guide vane adjustment mechanism for a Francis turbine

Photo 16 — A 35 kW Francis turbine with massive governor and 670 pound flywheel in the center

There are three main objections to the use of Francis turbines, especially on small installations. First, the runner is a single unit, and so if part of it is damaged sometimes the complete unit has to be replaced. Second, there is always the possibility of cavitation (see below). Third, very fine clearances are required for the moving parts, and if grit is suspended in the water it can destroy the turbine. A case in point is the following installation:

Figure 16 shows the general layout of an installation which used to produce 48 kW on a net head of 15 ft. and a flow of 3240 cfm. The Francis turbine, installed in 1930, ran happily at 300 rpm until a local mining group decided to dump their waste into the river a few miles upstream. First the reservoir was filled in to the height of the dam with grit, and shortly afterwards the runner was severely damaged, its output reduced from 48 kW to 1 kW.

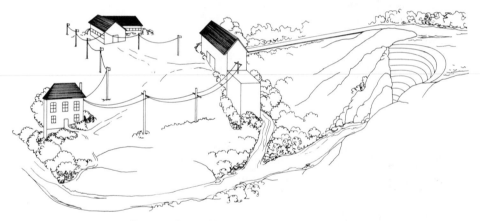

Fig. 16. General layout of a 48 kW installation

Another objection to Francis turbines is their cost. The owner of the above installation was quoted in excess of $36,000 for a replacement of the runner alone! The same engineering company also quoted $112,000 for new replacement hardware, to include a mechanical governor, alternator etc. A cross-flow turbine, either from Ossberger or one built locally to given specifications, would be incomparably cheaper, say about $16,000 for the turbine, electronic governor and alternator. Individually tailored Francis turbines continue to be very expensive. There are some manufacturers of standard turbine sets, described in the following pages, which use the Francis and propeller runners; these sets can in certain conditions work out to be quite economical. Some second-hand units can be found, but they are disastrous buys if the conditions they were manufactured for do not suit the head and flow characteristics of the new sites. Where possible, check with the manufacturer that the runner is suitable for its new site before buying it.

Cavitation

Cavitation is a rather peculiar disease to which propeller and Francis runners are prone under certain conditions. Basically, it is a result of tiny bubbles which form in the water and collapse when they come into contact with the runner blades. This causes a form of wear known as cavitation. It tends to occur when any of the following is too high: runner speed, water temperature and the altitude at which the turbine is situated. Cavitation is a problem for engineers manufacturing turbines, and when purchasing a Francis or propeller turbine one should ask the manufacturer what, if any, is the likelihood of cavitation occurring to the runner.

The Propeller Turbine

The propeller is an axial-flow high speed reaction turbine used on low heads, 3 to 30 ft. Precision engineering is required for the manufacture of the runner in order to attain its efficiency of 80% plus, and on small installations. The main disadvantage with the fixed blade propeller is its very low part-flow efficiency. Should the flow be reduced to 30% or less, the output will be nil. This handicap is overcome in large installations by using multiple runners and only operating, at full efficiency, as many as conditions will allow. Smaller installations can do likewise with two runners. A fixed blade propellor is a good investment if a constant flow of water is available.

One idea which might be useful on small installations is to have manually adjustable fixed runner blades. In spring, when the flow increases, the angle of the blades is adjusted to give a high efficiency and likewise when the flow decreases in summer.

Many megawatt installations employ this idea in what is known as a Kaplan turbine (see Figure 18). The blades of the Kaplan are automatically adjusted to suit flow conditions, but this degree of mechanical sophistication on a small installation would be impossibly expensive, unless it were manually operated. Viktor Kaplan, who designed the turbine, built the first one in 1919. The runner diameter was 2 ft. and under a head of 9.6 ft. it produced 26 kW with a high efficiency of 81% from 36 to 106% of the flow, a vast improvement on the propeller.

Like the Francis, both the propeller and Kaplan turbines are subject to cavitation (see above), but this tends to occur only under high heads or with excessive suction heads.

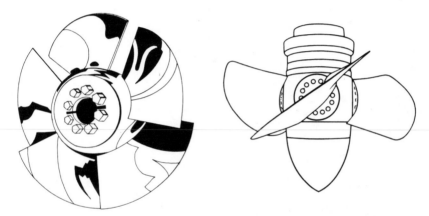

Fig. 17. Propeller turbine   Fig. 18. Kaplan turbine

Francis and Propeller Turbine Manufacturers

THE JAMES LEFFEL COMPANY,
SPRINGFIELD, OHIO 45501
(513) 323-6431
The Leffel Company has been manufacturing both large and small turbines for 114 years. Of particular interest is its standard range of Hoppes small hydro-electric sets.

Photo 17 — Hoppes unit designed to produce 5 kW at 120 volts; 60 hertz on a 25 foot head

Photo courtesy James Leffel & Co.

The Hoppes, on a head of 13 to 25 ft. and with a flow of 70 to 980 cfm, will produce between 0.5 and 10 KW. It is driven by a fixed-pitch propeller turbine made of bronze. The turbine shaft is made from heat-treated alloy steel. The generator sits above the turbine and is directly connected to it. A mechanical governor regulates the propeller speed by controlling flow of water through the unit. The electrical equipment and governor are protected by a steel plate housing. The switch panel has a main line switch, voltmeter, fused cutout and rheostat.

All in all the Hoppes is both a solid and well-made turbine. While Leffel does not want any general indication of prices given, it is happy to give a quotation upon receipt of site details. Their bulletin H49 gives this further data:

Head	Flow	Output
12 ft.	127 cfm	5 kW
13 ft.	470 cfm	6 kW
15 ft.	590 cfm	7.5 kW

From its many years of experience Leffel has built up hundreds of patterns and designs for propellers, Francis and Pelton turbines. It has the facilities and ability to build a turbine to suit any site and any purpose.

INDEPENDENT POWER DEVELOPERS, INC.,
P.O. BOX 1467
NOXON, MONTANA 59833
IPD's low-head fixed-pitch propeller turbine is a very welcome addition to the range of turbines currently available. With this unit, heads from 50 ft. down to as low as 5 ft. can be utilized to generate a steady 2 kW, given the right flow. The 32-volt DC generator is the same as that used with IPD's high-head Pelton turbine. The cost, including batteries and 3 kW inverter, is about $3,700.

G & A ASSOCIATES
223 KATONAH AVENUE
KATONAH, N.Y. 10536
(914) 232-8165
Back in 1972 Ken Grover started G & A, which specializes in small water power. Grover's interest in the subject began back on the farm where he was born, which was without electricity until he rigged a water-wheel on a nearby stream to turn a pair of truck generators, an inexpensive system that provided power for 20 years. With a degree in mechanical power engineering behind him, Grover's interest in water power continued. He has now invented, manufactured and installed a new radial flow turbine. While it is a Francis turbine, this new design incorporates some important cost-saving differences — which could make it attractive as a runner for use in many situations.

As well as the radial inflow, G & A also manufactures Kaplan turbines. Both are suitable for use on very low to medium heads. The average output of such turbines installed to date is 3 to 25 kW, but they may be scaled up to any size. Power generated is usually

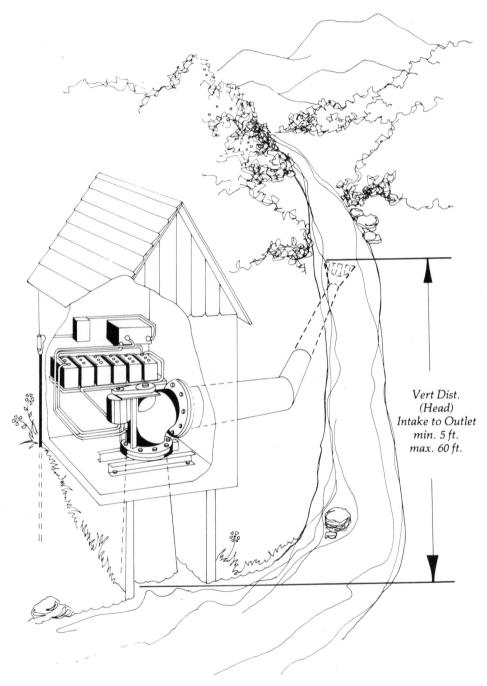

Vert Dist.
(Head)
Intake to Outlet
min. 5 ft.
max. 60 ft.

Fig. 19. Layout of IPD's low-head propeller turbine

used for heating, battery and inverter systems, or combinations of both. Send a dollar to G & A for further details.

BARBER HYDRAULIC TURBINES LTD.,
BARBER POINT,
P.O. BOX 340,
PORT COLBORNE, ONTARIO, CANADA L3K 5W1.
Barber has been manufacturing turbines since 1867, and currently it is developing a 'Mini-Hydel' unit in the test laboratory. The unit will be a semi-standard packaged set using propeller (for low heads) and Francis (high heads) runners. Base output is intended to be about 20 kW. Prices were not available on going to press.

An affiliated company, Canada Frontier Water and Power Limited, conducts feasibility studies, site visits, design of small dams and power plants, project management, etc.

ELEKTRO GMBH,
ST. GALLERSTRASSE 27,
WINTERTHUR, SWITZERLAND.
Elektro manufactures a small Francis turbine set. On a head of 25 to 64 ft. and with a flow of 32 to 64 cfm it will produce between 0.5 and 2 kW. The unit, which comes with a draft-tube and alternator but no governor, costs between S.Fr. 8000 and 10,000 ($3,200 and $4,000 approximately). It also produces Pelton turbines.

JYOTI LTD.,
R.C. DUTT ROAD,
BARODA — 390 005, INDIA.
Jyoti manufactures a range of three Micro-Hydel turbine sets as follows:

Output kW	Head Feet	Flow cfm	Cost
3-5	10-40	156-435	$5,500-8,000
10	20-40	281-543	$6,000-7,000
25	197-425	66-150	$7,500-12,000

All figures are approximations only. The standard unit includes mechanical governor (with flywheel), generator and control panel.

Francis and propellor runners are used on the low-head sets, Pelton and Turgo impulse wheels on the high-head version. Jyoti also manufacture a range of turbines with outputs from 25 to 1000 kW.

CANYON INDUSTRIES,
5346 MOSQUITO LAKE ROAD,
DEMING, WASHINGTON 98244

Canyon Industries is one of the few small turbine manufacturers in the world that use hydropower for their own energy needs (another is Land and Leisure Services). After two years of tests Canyon now produces the Hydromite (see photograph 18). On a head of 15 to 34 ft. and with a flow of 30 to 40 cfm it will produce 150 to 700 watts, if coupled to a suitable generator. The tiny output is bound to make megawatt men laugh, but I have quite some respect for this little turbine. A continuous 500 watt output will give 12 kW hours a day, not a lot but quite enough to meet the needs of a cottage for lighting and other small electrical appliances. But its main virtue is that it operates on a low head, whereas all other standard small output mini-turbines require a head of at least 50 ft. The unit, which weighs just 6½ lbs., costs $395, excluding the generator. Canyon hopes to have a unit complete with generator suitable for battery charging available soon. The runner is basically a propeller, with guide vanes to increase its efficiency. My only hope is that its light weight does not detract from its life-span. Further details are available from Canyon Industries. Send $1.00.

Photo 18 — The mighty Hydromite with alternator

LAND AND LEISURE (SERVICES) LTD.,
PRIORY LANE,
ST. THOMAS, LAUNCESTON,
CORNWALL, ENGLAND.

Land and Leisure is run by brothers Rupert and James Armstrong Evans, who carry on the hydro-power tradition from their father and grandfather. In fact the house where they were raised, and where they now have their workshop, is served by an old metal water-wheel and a first-class Francis turbine.

Land and Leisure supply new propeller turbines and Pelton wheels, either as complete units or just the runner alone. They are currently working on a number of installations mainly using twin sets of fixed-blade propeller turbines. On the twin-propeller installations, propellers of two different sizes are used. The large propeller will function at full efficiency on 70% of peak flow and the smaller one on 30%. When the flow is at its full peak both turbines are opened up. They find this method works out cheaper than a variable pitch Kaplan turbine. They are currently fitting a water-to-water heat pump (see page 82) on one such twin-propeller installation. Land and Leisure were the first in the water power field to develop and install electronic load governors.

K.M.W.   AND   A.B. BOFORS-NOHAB,
FACK, S 681 01,   FACK, S 461 01,
KRISTINEHAMN, SWEDEN.   TROLLHATTAN, SWEDEN.

Both these companies supply fixed- and adjustable-bladed propeller turbines with outputs from 100 kW upwards. The layout is shown in Figure 20.

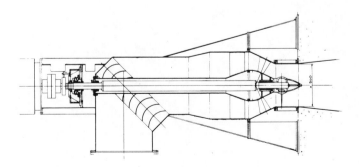

Fig. 20. Fixed blade, 100 kW propeller turbine

Dams

Dams are required for any or all of the following purposes:

(a) To divert the flow to the turbine or water-wheel

(b) To raise the head

(c) To create a storage reservoir

Diversion dams for small mountain streams are easily built to a height of 3 or 4 ft. Stones, logs, or other materials at hand can be used. All that is required is a small pool which will accept an intake pipe with a protective mesh at the opening. Many mountain streams form pools naturally, and in such cases there is little work to do. Stones or grit that gather in the pool can simply be shovelled out when necessary, rarely more than once a year.

Dams and reservoirs on large rivers designed to feed low-head installations are another matter altogether. In such cases a civil engineer or the local water authority should be consulted. It is unlikely that anyone reading this book would want to build a 750 ft. high dam, but one of that size was built in 1960, near Venice. The first time it was filled to capacity, the weight of the water triggered a massive rock slide and the resulting wave swept over the dam and killed more than 2000 people downstream. Even small on-stream reservoirs, when suddenly hit by flood waters, can cause considerable damage downstream. One way to avoid this is to build an off-stream reservoir.

Apart from the safety problems with reservoirs, they do tend to be very expensive, the land they drown can be costly, they inevitably silt up, and moreover, to make efficient use of the stored water, an expensive hydraulic governor should be used to regulate the flow to the turbine. Small reservoirs that can be used for enjoyment, fish farming, watering livestock, etc., are excellent, but anything bigger than that brings in legal complications and high costs.

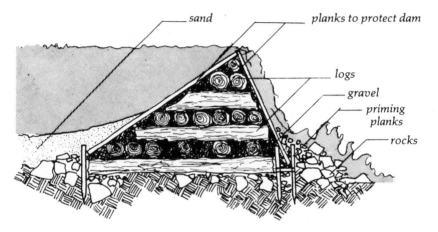

Fig. 21. Wooden dam

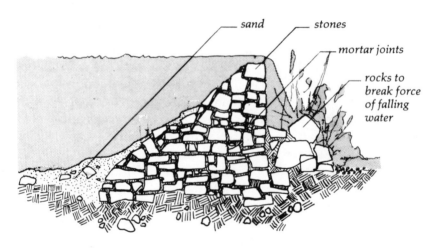

Fig. 22. Stone dam

For building small dams up to a height of 5 ft., local materials and designs can be the best. In most parts of the world there are bound to be other small dams or weirs in the locality. If possible find one on the stream or river you intend to dam, find out how it was built and — if you like it — build something similar. I know of a number of small dams in Ireland and Wales built entirely from cut peat or turf, and these, well trampled down, have stood the test of time. In Hungary, faggot dams can still be found, though they are not to be recommended. The more usual types of dams found today are earth fill, concrete gravity or crib dams built with logs, stones and clay or grit. Many such dams and dry-stone weirs have lasted hundreds of years without being washed away.

Before building any water impoundment permission often must be secured from local or state authorities. For the sake of safety and sturdiness I would urge any builder to seek experienced advice or consult one of the following books before constructing a dam:

SMALL EARTH DAMS
CIRCULAR 467 (23 pages),
CALIFORNIA AGRICULTURAL EXTENSION SERVICE
90 UNIVERSITY HALL
UNIVERSITY OF CALIFORNIA
BERKELEY, CALIFORNIA 94720

DESIGN OF SMALL DAMS
DEPARTMENT OF THE INTERIOR (816 pages),
SUPERINTENDENT OF DOCUMENTS,
U.S. GOVERNMENT PRINTING OFFICE
WASHINGTON, DC 20402

PONDS FOR WATER SUPPLY AND RECREATION.
Agriculture Handbook 387, available from US Government Printing Office, Washington, D.C. 20402, for $1.25.

Fish Passes

If a dam is built on a river frequented by such migratory fish as salmon or trout, a suitable fish pass or ladder must be constructed. It is only proper, as the fish have been using the river for a lot longer than we have.

Where a small dam is constructed, it is a simple matter to provide a series of pools with flowing water. Each pool should be about 18 inches or two feet higher than the previous one. If the dam is a high one, then larger resting pools should be provided at intervals so that the ascending fish can recover from their efforts. Another type of fish pass, known as the Denil after its inventor, consists of a steep, narrow channel with inclined baffles projecting from the bottom and sides. The baffles slow up the flow of water sufficiently to allow the fish to swim against it.

A problem arises when the young fish make their first voyage downstream, for if not protected they will be destroyed by the turbine. A fine mesh screen must be installed, and it is generally considered that the water velocity through these screens should not exceed one foot per second to prevent the fish being forced against the screens and killed. This can involve large areas of screens, which are easily clogged with debris, but it is worth the trouble to keep the river well stocked with fish. A screen is not required with overshot water-wheels (the young fish probably enjoy the ride).

Photo 19 — Waterfall with Denil fish pass

Trash Racks and Silt Traps

Trash racks or screens are required to prevent debris from entering the turbine and destroying it. They are made from flat metal bars and come in all shapes and sizes, depending on site conditions and the type of turbine in use. For example, a turbine on a river which can float logs is usually protected by two trash racks, one coarse and the other fine. Consult with the manufacturer of the turbine for the most suitable racks.

Should the river contain an excess of damaging silt, then a trap should be built. The silt trap or settling tank is usually placed before the trash rack and is simply a submerged section of the intake works which catches the silt. In most cases the trap needs to be of broad section to allow time for the suspended silt to settle. A by-wash must be provided to clean the trap and allow the silt to continue on its normal course.

Spillways and Sluices

A spillway must be built into every dam to allow the water which does not enter the turbine to flow on. It should be capable of accommodating the highest of flood-waters without damage to its structure, and without becoming clogged with branches or whatever. Spillways are also provided at the end of long leats or open channels.

Sluices, or sluice gates, regulate the flow of water to the turbine. They are situated at the intake or at the end of the channel and need to be sturdy enough to withstand flood waters. Usually, they are built of wood, grooved or notched, to allow the gate to slide up and down, and are operated through a rack and pinion geared to a crank through a worm and wheel.

Open Channels

Channels, sometimes called conduits, leats or canals (the engineering term is "headrace"), are used to carry water from a dam or weir direct to a water-wheel or to the pipeline feeding a high-head turbine. The quantity of pressure pipeline required in a high-head installation can be reduced by using a channel to carry the water to the point where the highest head is attained in the shortest possible fall. The question, to be resolved on the merits of each case, is whether the saving on pipeline cost can be justified against the cost of open channel construction.

Channels should, where possible, be excavated. Suspended channels, called flumes, can be used to carry water across ditches or other obstacles. The safe maximum velocity of water in channels is as follows:

Ordinary soil.	2-3 ft. per second
Ordinary good loam.	3-4.5 ft. per second
Best clay loam.	6-8 ft. per second

Higher velocities can be allowed in rock cuttings, or if the channel is of concrete, wood or steel. The principle is that the lower the velocity, the less head is lost, and the less trouble will be experienced with erosion of the sides.

Channels made with unlined earth may be subject to serious loss through seepage. The bed of the channels should be level for the first 100 yards or so, to allow any silt to settle. The channels should also incorporate a by-wash for washing out the silt every few years. The flow in the channels, after the first 100 yards, should be such that the channel is kept clear but is not eroded, and that is what dictates the gradient.

The design of channels (and dams) is the work of a civil engineer and I recommend that one be consulted before the construction of a large project.

Pipelines

Pipelines can be constructed from steel, iron, concrete, asbestos cement and plastic (PVC). In the U.S. and the U.K. for the present, PVC is the cheapest in most cases. Moreover, as the bore is of such fine quality, the head losses due to friction need not exceed 5% and should never exceed 10% — unless the layout of the pipeline goes up and down like a yo-yo. If the diameter of the pipe is too small for the flow there will be a serious loss of head and consequently power.

The best price is usually gotten directly from the manufacturer. Pipelines should be anchored securely, as there is the tendency for the whole line to slide down the hill. Remember that the greatest point of pressure is on the bends. Wherever possible it is worthwhile to bury the pipeline, because underground it is not an eyesore, blow-outs are less likely, damage due to sunlight is avoided and there is less likelihood of the pipeline freezing.

One idea for low-head large diameter pipelines is to use spiral wound galvanized steel tube of the type used for French drains and tunnel ventilation. They are considerably cheaper than PVC. My only worry would be their lifespan and reaction to being buried in clay. I know of no one who has tried the idea, but it is worth investigating.

Transmission Drives

Transmission drives may be classified as:
1. Belt drives (flat and V-belts)
2. Gear drives (bevel, spur and helical gears)
3. Chain drives
4. Motor vehicle gear boxes and back axles

Belt drives are most frequently used for small water-power plants. Torque is transmitted through a belt on the large driving pulley on the turbine shaft, to a small pulley on the generator shaft, thus stepping-up the speed of the drive from the turbine to suit the speed required by the generator. The ratio of transmission between a single pair or belt pulleys should not exceed 8:1. Flat belts, with an efficiency of 93-95% are now superseded by V-belts with efficiencies of 97% plus. The drive consists of multiple V-belts made of rubber with an inset of fabric. The belts ride in grooves in the pulleys and, due to the increased friction, they are more compact and efficient.

Bevel gears are used for driving horizontal-shaft generators off vertical-shaft turbines. Single-step bevel gears are suitable up to ratios of 7:1 with an efficiency of 98%. Whereas bevel gears enable one to turn a corner, spur gears connect in parallel. The highest ratio obtainable from a single-step spur gear is about 10:1, with an efficiency of 98%. The helical gear is an improved spur gear. Basically it has a longer life and makes less noise. More recently manufacturers have been introducing new types of gears, such as the elliptical gear which distributes the torque more evenly through a series of four wheels. The names and addresses of gear manufacturers can best be got from their advertisements in mechanical engineering journals.

Chain drives require less space than V-belts, but they tend to be more expensive. Properly designed and well lubricated they are noiseless in operation. Their ratio should not exceed 5:1. Efficiency can be as high as 99%.

The use of second-hand or new automobile differentials and gear-boxes is quite valid, provided enough is known about these forms of drive transmission. It is pointless, after rummaging around scrap-yards for months to find that the torque from the turbine will destroy the gear in a matter of months. However, I do know of one installation which, for the past two years, has successfully used an old truck back-axle with a ratio of 7:1 to take 10 kW at 200 rpm from a Francis turbine up to alternator speed. It should be noted that the owner of that installation is a mechanic, with a very well equipped workshop. The gear ratio for such second-hand parts can be found by turning one shaft and counting the resulting turns on the other shaft.

Alternators and Generators

All turbine manufacturers will know what type of alternator or generator is required to suit the particular turbines they supply. Indeed many of them have special arrangements with alternator manufacturers, based on a mutual understanding of the changes that may need to be made to the alternator before it can be used with a water turbine. I suggest that it is well worth purchasing the alternator through the turbine manufacturer or else to follow their advice as to the most appropriate alternator to use.

The problem is this. Most turbines have a runaway speed of about 1.8 times the normal full-load speed, and if the full load is suddenly removed, and the governor fails to act, it will run away in a matter of seconds. Standard alternators, usually driven by gasoline or diesel engines and therefore simple to govern, are only designed for a run-away factor of 1.2, whereas a water turbine has a runaway factor of 1.8. Thus if a turbine ran normally at 200 rpm it would run away to 360 rpm, and if the speed is increased in a 7.5:1 ratio, to drive a 1500 rpm generator, this speed would rise to 2700 rpm.

Put another way, the centrifugal force varies as the square of the speed. Thus if the force on the windings of a small generator is 1 lb, it will rise to 1.45 lbs. if the speed goes up to 120% of normal, but to 3.25 lbs. if it goes up to 180% of normal. If it is designed for a 100% safety factor at normal speed, it will be far beyond the safety limit at runaway.

The rotating parts of all turbines sold by experienced manufacturers are designed safely to withstand runaway speed, though the bearings may give trouble if the runaway is prolonged. The generator rotor is of much more delicate construction, with fine clearances between rotor and stator, and even if the rotor windings only stretch, they may touch the stator with disastrous results.

There are at least three ways in which this generator problem may be overcome. First, by installing a generator with an over-rated capacity, and thus capable of withstanding maximum over-speed. Second, by using an independently operated control device, such as an electro-magnetic clutch, which will automatically disconnect the generator from the turbine should the need arise. Third, by having a generator suitably modified to withstand the maximum overspeed.

Slow-speed alternators generally last longer than high-speed ones. Also the slower the speed the less transmission drive required. The more poles an alternator has, the slower the drive need be. Some examples of the number of poles and the speed required to generate at a 60 Hertz frequency are given below:

Number of Poles	Speed (rev./min.) for 60 Hertz output
2	3,600
4	1,800
6	1,200
8	900
10	720

There are few alternators around with more than four poles, but sometimes they can be picked up second-hand, and are often good buys.

Generators can be AC (alternating current) or DC (direct current). AC, produced by an alternator, can be single or three phase. All domestic appliances operate on single phase AC current, at 120 volts and a frequency of 60 Hertz. Three phase AC requires three circuits which must be balanced against each other.

DC, direct current, has a distinct advantage over AC in that no complicated governor is required for its generation. DC generators are used on low-output turbines and the current is fed to batteries. When power is required, it can be taken straight from the batteries through a DC circuit similar to that used in a car. Power can also be taken from the batteries through a solid state inverter, which will give AC current at whatever voltage is required. Alternatively it is possible, and indeed most economical, to feed the current directly into water heating or storage heaters, thus avoiding the governor and the battery/inverter system. Both of these arrangements are common to modern wind generator installations.

Another new and interesting way of avoiding costly governors and batteries is to interface a "Gemini synchronous inverter" between a power line and the turbine alternator. The Gemini inverter will instantaneously convert the varying frequency and voltage from the alternator to a constant 120 volts 60 Hertz. Moreover it will act as a battery (or mill-pond), as I have mentioned earlier, in that it enables excess power to be stored in the utility company's power line until it is required for use. Also should the turbine need to be shut down for servicing, or whatever, then normal power can be drawn from the line through the inverter. Thus one has the benefit of both worlds, but all this depends upon having a power line close by and upon the utility's approval. If for any reason the power line were closed, then a battery and inverter system could be used instead.

Windworks, which manufactures this inverter at Box 329, Route 3, Mukwonago, Wisconsin 53149, (414) 363-4408, reports that the utility companies tend to be reasonably co-operative in most states.

One other type of generator which can be used with small water turbines is the synchronous or induction generator. In truth, I don't know of a single installation using this type of generator, but its advantage is that it does not require a governor — in fact it serves exactly the same function as the Gemini inverter/alternator system, providing normal current when connected to a power line. But it has two outstanding disadvantages. First, it must be driven at constant speed, so if the flow of water decreases beyond a given point the generator will be useless. Second, if the power line fails, the induction generator is useless.

One AC alternator manufacturer is worth special mentioning:

LIMA ELECTRIC COMPANY, INC.,
200 EAST CHAPMAN ROAD,
BOX 918, LIMA, OHIO 45802,
(419) 227-7327

Lima produce a range of alternators which excel in two qualities. Take for example their 20 kW alternator: It is capable of starting a 20 HP motor and is 88.2% efficient. Moreover the Lima MAC alternator can run at 50% overload for one hour and at 25% overload for two hours.

Power Transmission

Voltage loss in the home hydro system will result from resistance in the wiring used. The resistance loss of copper and aluminum two-wire power transmission wire used between power generation source and the load, is shown in the table below:

Wire Gauge A.W.G.	Resistance Ohms per 100 ft. Two-Wire	
	Copper	Aluminum
000	.0124	.0202
00	.0156	.0256
0	.0196	.0322
2	.0312	.0512
4	.0498	.0816
6	.079	.1296
8	.1256	.206
10	.1998	.328
12	.3176	.522

The voltage drop in a wire is equal to amps times the resistance of the wire or: Volts = Amps × Resistance.

Power Loss in a wire is equal to amps squared times resistance or: Power Loss = $(\text{Amps})^2$ × Resistance.

Power Loss from the generation source to the load can be minimized for any size wire by increasing the generation voltage.

Example: *Power generated is 600 watts. 200 feet of No. 4 A.W.G. copper wire will be used. Wire resistance equals one tenth of an ohm or: Resistance = 0.1 ohm. Choice of generating voltage is 12, 24, 32, 120, 240, or 600.*

Power Generated (Watts)	Generator Voltage	Generator Amps Output	Line Voltage Drop (Volts)	Voltage at Load	Power Loss In Wire (Watts)	Power at Load (Watts)
600	12	50	5	7	250	350
600	24	25	2.5	21.5	62.5	537.5
600	32	18.75	. 1.875	30.125	35.15	564.85
600	120	5	.5	119.5	2.5	597.5
600	240	2.5	.25	239.75	.625	599.375
600	600	1	.1	599.9	.001	599.999

NOTE: *Doubling the generator voltage results in one quarter of the power loss and increasing the voltage 10 times to 120 volts results in one hundredth the power loss.*

Governors

A governor is required on all turbines (except those using Gemini inverters, DC or induction generators) to ensure that the electrical output agrees with the demand. The many ways of achieving this can be broadly classified as follows:
1. Mechanical flow governors
2. Electronic governors

Mechanical flow governors regulate the flow of water to the turbine so that the generator output agrees with the load. There are two ways of doing this. The first method, that of diverting the flow on and off the runner, is used on impulse turbines, the Pelton and the Turgo. It is a fairly inexpensive way of governing, but does have the disadvantage of wasting water, unless a spear valve is used.

The other method of flow governing, used on Francis, Kaplan and cross-flow turbines, is to regulate the quantity of water entering the turbine. There are many ways of achieving this, but most use some form of guide vane or wicket gate system. This invariably entails the use of expensive and complicated automatic equipment that can give the fine clearances required for accurate governing. A flywheel is also required to guarantee constant speed and frequency at changing loads. Moreover the pressure in the pipeline, (assuming there is one), frequently increases and decreases as the flow changes, and this can cause water-hammer and pipe blow-outs.

Electronic governors, as previously explained, regulate from the alternator rather than the turbine side. The electronic alternator governor, the first type, is really a voltage regulator similar to those used on automobiles and trucks to prevent the batteries from over-charging. Let us say, for example, that we have a turbine-driven alternator generating 5 kW, and suddenly the load (or demand) for electricity falls to 2 kW. Now there are 3 kilowatts buzzing around with nowhere to go and causing damage, just like a bunch of street corner kids. The alternator governor will sense this and instan-taneously alter the field excitation on the alternator, thus reducing output to suit the load. And that's it. This type of system is frequent-ly used with wind generator systems, too.

The second electronic governor system is the electronic load-diversion governor. With this type, the full output from the gener-ator is continuously used. It operates by diverting the electric out-put to and fro from primary needs to secondary needs. This diver-sion is achieved in an instant, and constant voltage and wattage is maintained.

The primary need is whatever the turbine was installed for, and the secondary need can be resistance, space, water or greenhouse heating, battery charging or whatever.

The load diverter can operate by monitoring the alternator fre-quency, using a phase lock loop (controlled by an oscillator). The excess output will then be diverted to a thyristor-controlled throw-away resistance or ballast circuit. Thyristors do cause mains inter-ference, and this has to be suppressed.

Electronic load governors are sophisticated devices and should be purchased only from the very few people who know about them, or designed by a competent electrical engineer.

Where either type of electronic governor is used, safety devices should be incorporated to disconnect the alternator in the event of overspeeding.

The following companies manufacture governors:

WOODWARD GOVERNOR COMPANY,
5001 NORTH 2ND STREET, ROCKFORD,
ILLINOIS 61101
Woodward Governor Company manufactures mechanical governors.

NATURAL POWER INC.,
NEW BOSTON, NH 03070
(603) 487-2456
Manufacture electronic alternator governors. One with a maximum capacity of 10 kW costs $475, and will require maintenance work in 10 years (about the same as a Woodward UG8).

Sad to say I know of no one in the U.S. who manufactures electronic load-diversion governors. Hopefully the situation will change soon, since the only people I know who do make them are in England. This is not much use since their voltage and frequency is different from that in the U.S. However, their address is:

LAND AND LEISURE SERVICES,
PRIORY LANE, ST. THOMAS,
LAUNCESTON, CORNWALL, ENGLAND.

Heat Pumps

To use a heat pump in conjunction with a water turbine, especially if the water flows near the house or factory, is eminently sensible. A good water-to-water or air heat pump will, for every kilowatt of power consumed, produce three or four kilowatts of heat. A heat pump is a device similar to a refrigerator, and it extracts low-grade heat from water, earth or air and transfers the concentrated heat to where it is needed. Heat pumps which extract heat from flowing water tend to have a higher efficiency than those which extract from air or earth.

My only worry is that if in years to come, millions of these were used to extract heat from water and transfer it to air, that the global heat balance may be upset. Certainly if hundreds of them were used on inland waterways then the temperature of those waterways would fall, which would be bound to affect their ecology.

It would be uneconomical and troublesome to install a heat pump in conjunction with a water turbine if the output from the turbine alone would satisfy one's heat and power demand. But take, for example, a situation where the output from the turbine is only 5 kW, yet 15 kW of heat is required, not an unusual requirement for many houses in winter. In such a case a heat pump can fill the energy gap. Number 44 of the magazine *Mother Earth News* ($2.00 from P.O. Box 70, Hendersonville, North Carolina 28739) contains two articles on heat pumps and a listing of manufacturers.

Surveying Equipment

Surveying equipment is available from the following:

BERGER INSTRUMENTS,
BOSTON, MASSACHUSETTS 02119

LIETZ,
892 COWAN ROAD,
BURLINGAME, CALIFORNIA 94010.
Professional surveying equipment is expensive and quite unnecessary for most small turbine installations. Both the above companies sell small and cheap hand-held sighting levels. Berger sends a free booklet on request entitled "How to use transits and levels for faster more accurate building."

THE LAWS

With the exception of England and Wales, the law throughout the world relating to hydropower is eminently sensible. As a natural, renewable source of energy its development is actively encouraged in most countries. It is frequently helpful to discuss any proposed project with local water and civil engineering authorities, as they can often make useful suggestions with regard to dam construction, flood levels etc.

Where your neighbor's land and your own is divided by the river, you will need his permission if you intend to divert the river. Be careful of flooding neighboring land if a dam is constructed. It can be highly economical to share the costs and the power of a turbine installation with your neighbor. Twice the output can be gained for a very small increase in the total cost. Moreover, with his cooperation, you will have additional help and no resistance to the idea. In the UK, the State has a monopoly on the sale of electricity and in the U.S., a tax is imposed on its sale. Thus a joint endeavor can be much more practicable than selling electricity.

The environmental impact of small hydropower systems is nil, and indeed, if they do have any effect they are likely to be beneficial, since rubbish is removed from the river and the water is oxygenated. The only disadvantage occurs where fish use a river that has been dammed to spawn, and a lack of consideration is given to their needs. (See Fish Passes.) It is wise, if not required, to consult the appropriate local agencies and cooperate with them in planning any stream impoundment.

U.S.A.

Since the laws regarding water vary from state to state, it would be impossible to deal with them fully without going into great detail. In general there is little resistance to small hydro-power. If a sizable dam or reservoir is required, an environmental impact statement must, in some cases, be sent to the Federal Department of Environmental Protection. There are many people in this country with turbines which do not interrupt the flow of water who just went ahead and built them of their own accord. Needless to say, there are no "abstraction" charges, though a tax is imposed on the sale of electricity.

Generally in the U.S., there are two types of water rights: Riparian rights and Appropriative rights. Riparian rights, originally brought through English common law, usually apply to the eastern and midwestern states where water is plentiful. Riparian rights are basically defined as follows: the owner of any land adjoining the river has a common law right to the 'reasonable use' of the natural flow and that this right is shared equally by all other properties along the river course. 'Reasonable use' means sharing the available water first for domestic use and then for irrigation, etc.

Appropriative rights evolved in the more arid regions of the country. In such places water that is used for other than domestic purposes must be claimed through a legal process to establish Appropriative rights. This means that the first persons to make a claim can use as much as they need. Those who follow can claim any water left over! Appropriative rights are the exclusive law in Montana, Idaho, Wyoming, Nevada, Utah, Colorado, Arizona, New Mexico and Alaska. The states bordering this dry area have a mixture of both Riparian and Appropriative rights and are Washington, Oregon, California, North and South Dakota, Nebraska, Kansas and Texas. Some older properties occasionally have these Appropriative rights included in the title deeds.

Water rights laws are some of the most involved around. Moreover, each state can have completely different laws from the next. The US Department of Agriculture can be helpful in providing information on water rights, dams and reservoirs.

Whatever the laws (which mainly relate to the consumption of water through irrigation or whatever) say, it should be remembered that a water-wheel or turbine does not drink water: all the water is returned to the river after use. The only real interference comes where the flow of water is interrupted because of a large dam, and who wants such a dam?

Canada

Generally speaking Canada's inland water resources are managed by the provincial governments. With the exception of Quebec, the provinces's legal systems are based on English common law and landowners accordingly, if not restricted by other statutes, have riparian rights which would allow an owner to make any use he might wish of "his" water, yet would hold him responsible for any detrimental effects he might cause downstream.

However, each province has enacted certain statutes which limit those rights, for the management and conservation of water for the common good. Quebec relies on a code of civil law which controls water development projects through the ownership of water-beds and by various other regulatory measures.

Each Canadian province has created a civil service agency to regulate hydroelectric power developments, and these agencies' powers vary depending on the province. Some of the legislation affect home hydro projects.

Federal laws affecting home hydro development pertain only in Canada's northern territories (the Yukon and Northwest Territories where all hydro plants must be licensed by the Minister of Indian & Northern Affairs), and where the waterway involved flows across the International Boundary with the United States. Some of these latter cases merely require licensing by the Canadian Minister of Fisheries and by Environment Canada (Ottawa), while others require joint Canadian-U.S. approval.

The federal Navigable Waters Protection Act might relate in some cases to small hydro developments, in that construction of a dam of causeway is prohibited in any waterway created or altered for navigation, except by license by the Minister of Transport (Ottawa). This Act also limits the stringing of power lines if they interfere with navigation.

For addresses of Provincial Offices, write Assistant Director General, Inland Waters Directorate, Fisheries & Environment Canada, Ottawa, Ont. K1A OE7 CANADA.

England and Wales

A painfully idiotic situation exists in England and Wales as a result of the Water Resources Act of 1963. It is unfortunate that when the Act was written no thought was given to hydro-power.

Within the next decade or two we will almost certainly have entered an era of energy scarcity, and with it restriction on small hydropower will have to go — that is, the State monopoly on the sale of electricity, and the imposition of water abstraction charges by the water authorities. Until 1964 many small turbines in England and Wales supplied electricity to whole villages or towns but then they were closed down because it was made illegal for anyone other than the State to sell it. So, since 1964 they have been importing oil and burning non-renewable coal at an efficiency of 25% instead of continuing to use a perfectly good renewable source. This monopoly and the attitude of the Water Authorities in England and Wales will both have to change — and the sooner the better.

Scotland

Fortunately, the attitude of the Scottish authorities is a sane one since there are no restrictions on the development of water power with outputs of up to 50 kW, nor is there any nonsensical "abstraction" charges whatever the output.

Section 22 of the Hydro-Electric Development (Scotland) Act 1943 requires that the consent of the Secretary of State for Scotland must be sought before building a private hydroelectric station should its output exceed 50 kW. In practice it is unlikely that there would be any objection to the building of such stations. Indeed there are a number of cases where the North of Scotland Hydro-Electric Board assist such stations by purchasing their surplus output.

The construction of dams and reservoirs to impound in excess of 800,000 cubic feet are governed by the Reservoirs Act 1975. The Common Law rights of riparian owners downstream from the plant must be respected, in other words the flow of water downstream should not be meddled with. The North of Scotland Hydro-Electric Board at 16, Rothesay Terrace, Edinburgh EH3 7SE will be able to advise further.

Warning

Great care should be taken in the purchase of second-hand turbines and water-wheels, for each is designed to serve a specific head and flow. Therefore, they will not be suitable for use in different conditions. There was quite a bit of dealing in second-hand turbines after the last war, and some people made dreadful mistakes. All the blame does not rest with the scrap merchants, as they know nothing about what they are selling. They will buy, for example, an old 10 kW Francis turbine in good working condition and offer it for sale as such. What they do not explain is that this turbine had given 10 kW with a net head of 30 ft., and the person who now bought it expected 10 kW with a gross head of 13 ft. at an old mill. Can you imagine the disappointment, after paying good money for the turbine and having spent weeks of hard work installing it, to find that the installation was useless?

The only way to avoid such a disappointment is to write to the manufacturer, giving the turbine serial number and address of the original site, and ask if it is suitable for your needs. The manufacturer may give this information free but do not expect him to. You may ask at the same time if they are willing to recondition the turbine. If you cannot contact the turbine manufacturer in question and if your knowledge of water power is not very deep, then I strongly advise that you look elsewhere for a turbine.

Small Water Power Directory

America

Barber Hydraulic Turbines, Ltd.,
Barber Point,
P.O. Box 340,
Port Colborne,
Ontario, Canada L3k 5W1.
See Francis and propeller turbine manufacturers.

Campbell Water Wheel Company,
420 South 42nd Street,
Philadelphia, Pennsylvania 19104.
See Water-wheel section.

Canyon Industries,
5346 Mosquito Lake Road,
Deming, Washington 98244.
See Francis and propeller turbine manufacturers section.

Cumberland General Store,
Route 3, Box 479,
Crossville, Tennessee 38555.
Market the Leffel Hoppes unit.

G & A Associates,
223 Katonah Avenue,
Katonah, New York 10536.
See Francis and propeller turbine manufacturers.

GUY IMMEGA,
LASQUETI ISLAND,
BRITISH COLUMBIA, CANADA.
Willing to act as a consultant on small hydropower systems in his
area, has written for a number of books and journals on the subject.

INDEPENDENT ENERGY SYSTEMS, INC.,
P.O. BOX 1265,
BLOWING ROCK,
NORTH CAROLINA 28605.
Apart from installing turbines manufactured by Leffel and Inde-
pendent Power Developers, I.E.S. are engaged in development
work on their own complete system and on a small impulse turbine.

INDEPENDENT POWER DEVELOPERS,
BOX 1467,
NOXON, MONTANA 59853.
For product range, see Pelton and propeller manufacturers.

JAMES LEFFEL AND COMPANY,
SPRINGFIELD, OHIO 45501.
For product range, see Francis and propeller turbine manufacturers.

LIMA ELECTRIC CO., INC.,
200 EAST CHAPMAN ROAD,
BOX 918,
LIMA, OHIO 45802.
See under Alternators and Generators.

NIAGARA WATER WHEELS LTD.,
BOX 326, BRIDGE STATION,
NIAGARA FALLS, NEW YORK 14305.
Handle small to medium sized turbines. They offer design, manu-
facturing and installation services using new and rebuilt turbines.

SMALL HYDROELECTRIC SYSTEMS AND EQUIPMENT,
P.O. BOX 124,
CUSTER, WASHINGTON 98240.
For product range see Pelton wheel section.

F.W.E. STAPENHORST, INC.,
285 LABROSSE AVENUE,
POINTE CLAIRE,
QUEBEC H9R 1A3, CANADA.
North American representative for Ossberger turbines.

Europe

A.B. BOFORS-NOHAB,
FACK, S46101
TROLLHATTAN, SWEDEN.
See Francis and propeller turbine manufacturers.

VINCENT ALLEN ASSOCIATES,
291 HIGH STREET,
EPPING, ESSEX, ENGLAND.
Consulting engineers with an interest in small water power.

P.W. AGNEW,
DEPARTMENT OF MECHANICAL ENGINEERING,
THE UNIVERSITY,
GLASGOW, SCOTLAND.
Will advise on equipment etc. in Scotland. Developing a small propeller turbine.

DREES AND COMPANY,
4760 WERL/WEST,
POSTFACH 43, WEST GERMANY.
No longer manufacturing standard sets.

C. DUMONT AND CIE.,
PONT DE STE. UZE 26240,
ST. VALLIER,
DROME, FRANCE.
Manufacture small Francis, Pelton and Kaplan turbines together with all associated equipment.

ELEKTRO GMBH,
ST. GALLERSTRASSE 27,
WINTERTHUR, SWITZERLAND.
For product range, see Pelton and Francis sections.

ESCHER WYSS LIMITED,
CH-8023 ZURICH, SWITZERLAND.
Have just started production of 'Mini-Straflo' turbines. With outputs starting at 400 kW, it is not exactly mini.

GILBERT GILKES AND GORDON LIMITED,
KENDAL, WESTMORLAND, ENGLAND.
See Turgo impulse turbine section.

INTERMEDIATE TECHNOLOGY DEVELOPMENT GROUP LIMITED,
9 KING STREET,
LONDON WC2E 8HN.
Mainly involved in the development of low-head propeller turbines and high-head Pelton wheels in the 5 to 50 kW capacity.

K.M.W.
FACK, S681 01
KRISTINEHAMN, SWEDEN.
See Francis and propeller turbine manufacturers.

LAND AND LEISURE (SERVICES) LIMITED,
PRIORY LANE,
ST. THOMAS, LAUNCESTON,
CORNWALL, ENGLAND.
See Francis and propeller turbine manufacturers section.

OFFICINE BUEHLER,
TAVERNE,
CANTON TICINO, SWITZERLAND.
Reported to manufacture a wide range of turbines.

OSSBERGER TURBINENFABRIK,
8832 WEISSENBURG,
BAYERN, GERMANY.
See cross-flow manufacturers section.

WESTWARD MOULDINGS LTD.,
GREENHILL WORKS,
DELAWARE ROAD,
GUNNISLAKE,
CORNWALL, ENGLAND.
See water-wheel section.

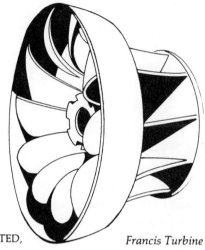

Francis Turbine

Other Countries

BALAJU YANTRA SHALA (P) LIMITED,
BALAJU, KATMANDU, NEPAL.
See Cross-flow section.

BARATA METALWORKS AND ENGINEERING P.T.
J.L. NGAGEL 109,
SURABAYA, INDONESIA.
Manufacture Francis and Cross-flow turbines with outputs ranging
from 18 kW upwards.

JYOTI LIMITED,
R.C. DUTT ROAD,
BARODA 390 005,
INDIA.
See Francis and propeller turbine manufacturers.

LOW IMPACT TECHNOLOGY (AUSTRALIA)
34 MARTIN STREET,
SOUTH MELBOURNE,
VICTORIA, AUSTRALIA.
Alain Gerrard, who started L.I.T., powers his home with a water
turbine. Can supply and install a whole range of natural energy
equipment.

SPEEDRIGHT EQUIPMENT LIMITED,
P.O. BOX 169,
LEVIN, NEW ZEALAND.
Manufacture a small turbine set for battery charging.

TIENTSIN ELECTRO-DRIVING RESEARCH INSTITUTE,
TIENTSIN, CHINA.
Trial-producing a range of small turbines, 0.6 to 12 kW capacity.

The above list gives the names of all those who have a special interest in the manufacture and installation of small water power plants. No doubt there are many small engineering firms spread throughout the world which have, in the past, made small turbines and would, if asked, be willing to manufacture to order. It is worthwhile asking in your locality if there is anyone who owns a turbine and finding out who the manufacturer was.

The journal, "International Water Power and Dam Construction" publishes a directory of suppliers and manufacturers of large-scale water turbines and associated equipment.

The author would appreciate receiving details of any new water turbines or manufacturers. Such information would be included in the next edition of this book. Please write with the information to Dermot McGuigan, c/o Garden Way Publishing, Charlotte, VT 05445.

Bibliography

THE BANKI WATER TURBINE.
See Cross-flow bibliography.

B.Y.S. CROSS-FLOW TURBINES.
See Cross-flow bibliography.

CENTRIFUGAL PUMPS, TURBINES AND PROPELLERS.
William Spannhake, M.I.T. Press (USA) 1934. Highly technical.

THE CROSS-FLOW TURBINE.
See Cross-flow bibliography.

DESIGN OF SMALL DAMS.
See under Dams.

A DESIGN MANUAL FOR WATER WHEELS.
See Water-wheel bibliography.

ELECTRICITY FOR THE FARM.
F. Anderson. Macmillan Co. (USA) 1915. Subtitled 'Light, heat and power by inexpensive methods from the water-wheel or farm engine.' I enjoyed it. Good for those who want to use a water-wheel with a DC generator.

ELECTRIC POWER PLANT INTERNATIONAL.
Published yearly by The Electrical Research Association, Cleeve Road, Leatherhead, Surrey, England. It includes the names and addresses of all alternator and inverter manufacturers in the world, together with details of their range. Very expensive, try your local or college library.

ENERGY PRIMER. PORTOLA INSTITUTE (USA).
  1974 and Prism Press (UK) 1976. Contains a good section on
  small water power. At the time of writing a new edition is in
  preparation and should appear late in 1977.

ENGLISH WATERMILLS.
  L. Syson. Batsford (UK) 1965.

FARM WATER POWER.
  G. Warren. US Department of Agriculture 1931.

HANDBOOK OF HOMEMADE POWER.
  Bantam Books (USA) 1974.
  It contains a reprint of a five-part article published in 1947 by
  Popular Science, and includes plans for a five foot diameter water-
  wheel with an output of between 350 and 700 watts.

HYDRAULIC ENGINEERING.
  Turneaure and Black. American School of Correspondence 1909.
  Dated and tending towards large installations.

HYDRAULIC SYSTEMS AND MAINTENANCE.
  Bradbury. Iliffe (UK) 1972.

HYDRAULIC TURBINES.
  Miroslave and Nechleba. Prague.
  Good on the mechanical design of turbines.

HYDROPOWER.
  A MacKillop. Wadebridge Ecological Centre, Cornwall, England.
  Half of its 74 pages has nothing to do with hydropower. The
  rest of it, apart from the reprint of the 1949 Popular Science
  article, is of little use.

HYDRO POWER ENGINEERING.
  James J. Doland. Roland Press Co. (New York) 1954.
  A textbook for civil engineers.

INTERNATIONAL WATER POWER AND DAM CONSTRUCTION.
I.P.C. (New York and London).
The journal for the water power trade, very good but rarely has anything on small turbines.

LOW-COST DEVELOPMENT OF SMALL WATER-POWER SITES.
See cross-flow bibliography.

NATURAL SOURCES OF POWER.
R.S. Ball. Constable (London) 1908.
Wind and water power, but mainly water.

PONDS FOR WATER SUPPLY AND RECREATION.
See under dams.

POWER DEVELOPMENT OF SMALL STREAMS.
Harris and Rice. Rodney Hunt Machine Company, Orange, Massachusetts, 1920.

POWER FROM WATER.
T.A.L. Paton and J. Guthrie Brown. Leonard Hill Ltd. (London) 1961.
Good general introduction to large-scale hydro-electric schemes, mainly in the UK.

PRACTICAL WATER POWER ENGINEERING.
W.T. Taylor. Van Nostrand (USA) 1925.
and Crosby Lockwood (London) 1925.
Deals with rainfall and run-off, measurement of flow. Good on reservoirs, site selection and canals. 60 pages on electric power transmission! Small to medium installations. Misleading title.

SENSITIVE CHAOS.
T. Schwenk. R. Steiner Press 1965.
An excellent, if rather esoteric, photographic and written study of water and wind as living, flowing elements. Recommended.

SMALL EARTH DAMS.
See under Dams.

SMALL SCALE POWER GENERATION.
   United Nations 1967.
   Wind and Water Power.

THE SOCIETY FOR THE PROTECTION OF ANCIENT BUILDINGS.
   Has published a series of booklets on European watermills. Write
   to S.P.A.B., 55 Great Ormond Street, London WC1.

TRANSACTIONS OF WORLD POWER CONFERENCE.
   1924, Volume 2, Water.
   The Central Office of the World Energy Conference,
   201 Grand Buildings, Trafalgar Square,
   London WC1.
   814 pages on water, mainly big plants but some small.

TREATISE ON MILLS AND MILLWORK.
   Sir William Fairbairn.
   Longmans (London) 1861, 1878.
   The complete work on watermills. Available through some library
   services.

WATER POWER DEVELOPMENT.
   E. Mosonyi. Hungarian Academy of Sciences. Budapest 1960.
   Vol. 1. Low-Head Water Plants. Vol. 2. High-Head Water Plants.
   Includes an interesting section on 'Midget power plants.'

WATER POWER ENGINEERING.
   R. Hammond (London) 1958.
   Megawatt plants.

WATER TURBINE PLANT.
   J.O. Boving, Raithby, Lawrence and Co. 1910.
   Lots of good drawings and technical information on Pelton and
   Francis turbines.

WATER TURBINES.
   P.M. Wilson. H.M.S.O. (London) 1974.
   Very good introductory booklet.

Turbine Efficiency Graphs

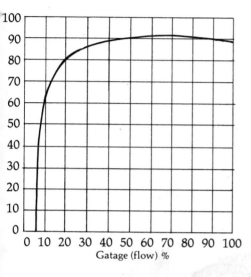

Efficiency curve for high head impulse turbines, Pelton and Turgo.

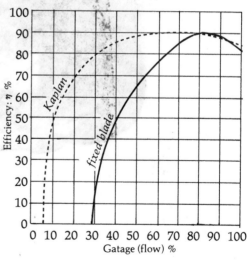

Efficiency curve for low head, single cell cross-flow turbines.

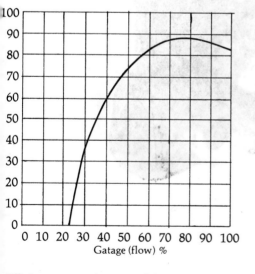

Efficiency curve for typical low head Francis turbine.

Efficiency curves for low head fixed blade propeller and Kaplan turbines.